LEVEL UP WITH AI

LEVEL UP

WITH

Ai

A Beginner's Guide to Safe, Productive, and Ethical AI
for Work, School, and Small Business

K. P. SEYMOUR

TABLE OF CONTENTS

INTRODUCTION

On my first day teaching medical students, I remember looking out at a lecture hall filled with eager faces and realizing the daunting task before me. My job was not just to transmit messages; it was to prepare these students to care for real human beings. That meant two different kinds of training: the preclinical foundation, which includes vocabulary, physiology, and numerous diagrams that form the basis of medicine; and the clinical application, where theory meets the messy, unpredictable reality of patient care.

For years, that balance was struck through traditional lectures, lab sessions, and hands-on clinical rotations. However, in recent years, a new dimension has quietly emerged: artificial intelligence (AI). At first, it appeared on the edges of everyday appliances, embedded in online modules that helped students test themselves, or in simulations that allowed them to "interview" digital patients who never grew tired, forgot their backstory, and never got offended when a question was awkwardly phrased. Now, with the arrival of AI avatars and conversational models, I can see the outlines of something transformative: a way of teaching and learning that is more personalized, more interactive, and perhaps more human than before.

And yet, outside of medical schools or research labs, whenever I bring up AI, the reactions are nearly always the same: nervous laughter, skepticism, or a confession of fear. "I am not a tech person," "That stuff is too complicated," or "Isn't AI dangerous?" These are not unreasonable concerns. For many people, AI feels like a black box, a force of nature that belongs to Silicon Valley engineers or sci-fi movies, not to ordinary, everyday people.

Here is what most people do not know: AI is not magic, and it is not reserved for the technically gifted. It is a set of surprisingly approachable tools that can already help you write, learn, create, and solve problems more efficiently. You do not need to code or have a computer science degree. What you need is the same curiosity or adaptability you have used to learn anything new in your life: from figuring out how to use email for the first time, to navigating a smartphone, to adjusting to a new job or family role.

This book is written for people who have watched the AI conversation explode and thought, *This is not for me.* I want to show you why this is not true. If my medical students, some of whom arrive with minimal technical skills, can use AI tools to learn complex diseases, practice patient interviews, and sharpen their clinical judgment, you can use AI to simplify tasks in your own work and daily life.

Why, then, does AI feel so intimidating? Part of it is the language. Words like "machine learning," "neural networks," and "large language models" sound alien, almost mechanical. Part of it is the hype. News headlines swing between extremes: AI will either save or destroy us. Part of these beliefs is cultural memory: Many of us grew up with stories in which machines become our rivals, replacing human workers or even turning against their creators.

The reality is less dramatic and far more interesting. AI is neither savior nor villain. It is more like a capable assistant: helpful, sometimes brilliant, occasionally clumsy, always dependent on human guidance. If you learn how to use it wisely, it can save you time, spark your creativity, and even reduce the mental load of daily life. If you ignore it, you may not be immediately harmed, but you risk missing out on one of the most accessible revolutions in technology since the internet.

I know what it feels like to hesitate. Even as a professor surrounded by digital tools, I initially kept AI at arm's length. I told myself I didn't have time to learn, that my students didn't need it, that the old ways worked well enough. But then I began experimenting: first with drafting lecture outlines, then with building interactive modules, and finally creating patient avatars that could carry on a realistic conversation. What surprised me most was not the power of the technology; it was how natural it felt once I let go of my skepticism. AI did not replace me as a teacher; it amplified me. It gave me more focus on the human side of teaching: the coaching, mentoring, and personal encouragement that no machine can provide.

That is what I want for you as a reader. Not to become an expert in algorithms, not to be dazzled by jargon, but to feel confident enough to say: "I can use this. I can make this work for me." Whether you are a small business owner juggling too many responsibilities, a professional trying to stay sharp in a changing workplace, or a lifelong learner who simply wants to understand what all the fuss is about, this book is your entry point.

In the chapters ahead, we will start simple. First, we will strip away the myths and confusion around AI and show you, in plain language, what it actually is. We will explore how you can use AI for everyday productivity, like organizing your calendar, writing better emails, or brainstorming new ideas. We will look at AI's role in work and business, and how it can help teams collaborate, streamline operations, and even spark new side hustles. And since technology is never neutral, we will talk about ethics: how to use AI responsibly, protect your privacy, and keep human values at the center of innovation.

This is not a textbook or a manifesto; it is a hands-on guide, filled with examples, exercises, and practical tips. My hope is that by the time you finish, you will be able to do more than understand AI. You will already have experimented with it and woven it into your daily life routine in ways that make your life easier, not harder.

Think back to the first time you used a smartphone: It probably seemed overwhelming at first, but within weeks, you were sending messages, taking photos, and downloading apps without a second thought. AI is at that same threshold. Today, it may feel strange, but soon, it will be as ordinary as checking your email or searching the web. The only question is whether you will approach it with fear or with curiosity.

PART I:
UNDERSTANDING AI

CHAPTER 1:

WHAT AI REALLY IS— AND WHAT IT ISN'T

EVERY FEW DECADES, A NEW INVENTION ARRIVES THAT changes the way people live and work in a way that is impossible to ignore. Think about electricity in the late 19th century, the personal computer in the 1980s, or the rise of the internet in the 1990s. Each of these breakthroughs started as something mysterious and intimidating, something only experts understood, but eventually they seeped into everyday life until no one could imagine living without them. AI is that kind of technology today.

If it feels like AI is everywhere right now, that is because it is. You have likely seen the headlines claiming that the technology will revolutionize work, replace jobs, cure diseases, or even threaten humanity. Politicians are debating it. Companies are racing to adopt it. Consequently, schools are rethinking how they teach. It has also found its way into everyday conversations in the office, school, and at the dinner table. People are asking the same questions: "What exactly is AI?" "How will it affect me?" and "Should I be worried or excited?"

Part of the reason why everyone is talking about AI is simple visibility. Tools like ChatGPT, Midjourney, and other AI-powered platforms have made advanced technology suddenly accessible to ordinary people. Now, you no longer need to be a computer scientist or a software engineer to experience what AI can do. You can type a question in plain English, or describe an image you want, and within seconds, the system responds in a manner that seems almost magical. That ease of access is a major shift. It took years for computers and the internet to move from laboratories into people's homes, and it seems as if AI made that leap overnight.

Another reason why AI is on everyone's lips is versatility. Unlike past technologies that tended to have a single function, such as a car for transportation or a phone for communication, AI is emerging in multiple domains simultaneously. It helps doctors analyze medical scans, assists lawyers in drafting contracts, guides teachers in creating personalized lesson plans, and enables entrepreneurs to brainstorm new business ideas in minutes. It is being used in agriculture to monitor crops, in finance to detect fraud, and in entertainment to generate scripts and music. No matter who you are or what you do, there is a good chance AI will touch some part of your work or personal life in the near future.

There is also a cultural impact that must be considered, which includes the fascination (and sometimes the fear) that machines can "think." Science fiction has fueled these feelings for decades, from stories of helpful robots to dystopian tales of rogue machines. When people hear about AI today, it often triggers those deep-seated cultural memories. That is why the conversations swing between extremes: utopian visions of AI solving humanity's biggest problems, and apocalyptic warnings about losing control to technology. The truth, as usual, lies somewhere in between.

Finally, the pace of AI development itself is fueling the buzz. Unlike many technologies that evolve gradually, AI is advancing at a breathtaking speed. Systems that seemed impossible a few years ago are already in use today. That rapid progress creates excitement and anxiety where people do not want to be left behind, but they are also unsure of how to keep up.

Therefore, when you hear people talking about it in the news, in workplaces, in schools, or even in casual chats, remember it is not just the hype. It is the recognition that we are living through a turning point, a moment when new tools are moving from the margins to the mainstream. Just like it happened with electricity, computers, or the internet, the question is: How will *you* choose to use AI?

WHAT IS AI?

By now, you have probably heard the phrase "artificial intelligence" so many times that it almost feels like a buzzword. Before you can understand what AI can do for us, we need to cut through the jargon and define it in plain, everyday language. AI is the ability of a machine or a computer to perform tasks that are usually associated with human intelligence. That could mean recognizing speech, understanding language, identifying objects in a picture, solving problems, or even making decisions. If a machine is performing a task that normally requires human thinking, it falls under the umbrella of AI.

At the same time, although it sounds straightforward, it can get confusing quickly, because AI is really an umbrella term. It actually covers a range of different techniques and technologies, and other phrases like *machine learning* (ML) or *deep learning* (DL) will pop up, making it seem like a blur of technical language. Here is a small breakdown:

- **ML** is a subset of AI. Instead of being programmed with strict rules for every situation, an ML system "learns" from data. For example, if you show a program thousands of pictures of seals and whales, along with the correct labels, it eventually figures out how to tell the difference on its own. ML does not mean that the machine is thinking; rather, it's more like pattern recognition at scale.

- **DL** is a further subset within ML. It is inspired by the way our brains work, using structures called "neural networks." DL stands out at learning complex tasks like recognizing faces, translating languages, or generating humanlike text. The voice assistants on your phone, for example, or the recommendation systems on Netflix and YouTube, rely heavily on DL.

To keep it simple:

- **AI** is the broad concept: machines doing seemingly intelligent things.

- **ML** is one way to build AI: teaching machines with data.

- **DL** is a specific technique within ML: using layered neural networks to solve really complex problems.

However, even this breakdown can give the wrong impression if we are not careful. People sometimes imagine AI as a single thing: a brainlike machine that either exists or does not. In reality, AI is more like a spectrum. On one end, you have narrow or "weak" AI systems that do one thing really well, like recommending a song or filtering spam from your email. On the other end, you have the idea of "general" AI, a system that could think and reason across several domains the way a human does. That general vision does not exist yet, and may not for a very long time. Almost everything you see today, from chatbots to image generators, lives firmly in the "narrow AI" category.

The spectrum idea is important because it helps us stay realistic. The AI you will use in your everyday life is not a sci-fi robot that can outsmart you. It is a set of effective and specific tools. To help you think about how AI can adapt to your life, think about it as a capable intern you have just hired to help you.

A good intern can be incredibly useful. They can draft documents, organize information, answer basic questions, or even offer you creative ideas. Sometimes, they will even surprise you with how good their work is. At the same time, they are not a seasoned professional; they definitely do not know everything. They will make mistakes, occasionally misinterpret instructions, and they work best when given clear guidance. If you leave them completely unsupervised, things can go wrong.

When you stop to see what AI can do, you will realize that it is not a robot plotting its next move; it is like an assistant who is fast, tireless, willing to help, but who depends on your direction. The better you are at giving instructions (often called prompts), the better AI can serve you.

Understanding this analogy also removes some of the fear of AI. Instead of imagining a machine that might replace you, think of it as one that can support you. Just as an intern allows a manager to focus on higher-level strategy, AI can free you up to spend more time on creative, human, or complex tasks that machines cannot handle.

Therefore, when you hear the terms "artificial intelligence" and "AI," remember that it does not mean a computer suddenly woke up and became self-aware. It means a collection of tools that excel at following patterns, learning from data, and giving you useful results as long as you remain in charge.

AI VS. AUTOMATION VS. ALGORITHMS

When people discuss AI, they often mix it up with other terms like "automation" and "algorithms." The confusion makes sense as these concepts overlap, but they are not the same. Understanding the differences clears up much of the mystery around AI:

- An **algorithm** is simply a set of instructions or a recipe for solving a problem. You already use them daily. A cookie recipe is an algorithm: Follow steps A, B, and C, and you will get the same dish. Computers do the same thing. A calculator, for example, follows an algorithm to multiply 7x9. It is not actually thinking, but executing rules. Algorithms are the foundation of everything in computing. One example is a navigation app calculating the shortest route to the grocery store. It runs the step-by-step rules to compare routes and pick the shortest.

- **Automation** is the use of technology to complete a process with little or no human input. This is a process about speed and efficiency, just as an automatic car wash. Once you press the start button, the system runs the same sequence every time. In offices, payroll systems that deposit salaries or email filters that remove spam are examples of automation. They do not adapt; they just repeat. Your car's cruise control is one application of automation. Once you set the speed, the system keeps it steady automatically—no thinking involved.

- **AI** is different because it adapts and learns from data. While automation repeats tasks, AI can recognize patterns and adjust. Instead of a spam filter that blocks every email with the word "lottery," an AI filter notices the kinds of messages you personally flag as spam and improves its accuracy over time. You will see this in modern navigation apps that predict traffic, reroute you if there is an accident, and estimate your arrival based on real-time driving data. That is AI adjusting to new information.

An easy way to remember is that the algorithms are the recipes, automation is the kitchen appliance that automatically runs the recipe, and the AI is the assistant who watches you cook, learns your tastes, and says, "Want me to add that cinnamon again like last time?"

The crucial difference is actually adaptability. Algorithms do not adapt; they only execute. Just as automation does not adapt, it repeats. AI can notice patterns, learn from experience, and improve performance. This matters because it helps us see through the hype. Not every tool marketed as AI truly qualifies. Sometimes, it is just clever automation wrapped in buzzwords. A quick way to tell the difference is to ask, *Is this system following fixed rules or is it actually learning?*

Together, automation, algorithms, and AI form a toolbox. Each is valuable, but only AI brings flexibility and learning into the mix. That is why it feels so different and why people cannot stop talking about it.

TYPES OF AI YOU WILL ENCOUNTER

One reason why AI feels overwhelming is that it does not just come in one form. Instead, there are numerous AI tools, each built for specific tasks. The good news is you do not need to master them all. Once you understand the major categories, you will be able to spot them in your daily life and know how useful they can be.

CHATBOTS

When most people think of AI today, they think of chatbots, which are programs that can understand and respond to human language. Tools like ChatGPT, Claude, Gemini, and Microsoft Copilot are examples. They allow you to type a question or request in plain English and get back a response that reads like it came from another person.

- *Use cases:* You can ask a chatbot to summarize an article, rewrite an email in a friendlier tone, brainstorm ideas for a project, or even simplify a complex topic. Think of them as conversation partners that are fast, knowledgeable, and endlessly patient.

IMAGE GENERATORS

Another popular AI category is image creation. DALL·E and Canva Magic Studio are examples. In this case, you type in a description, and the tool generates a custom picture in seconds.

- *Use cases:* If you need visuals for a presentation, social media post, or a personal project, image generators can save hours of searching. They are also great for sparking creativity: Even if the first image is not perfect, it often inspires new directions.

AUDIO AND VIDEO TOOLS

AI is not limited to just text and pictures, especially since it is also transforming the sound and video industries. Tools like Descript let you edit audio by editing text, and fixing an audio mistake is as simple as deleting a word. Eleven Labs can clone voices or create lifelike narration from text. Video platforms are emerging that can generate short clips, automatically edit scenes, and even create avatars that speak multiple languages.

- *Use cases:* Podcasters and video content creators can clean up recordings without technical skills. Businesses can generate training videos without hiring a voice editor. A student can record a presentation and use AI to polish the delivery.

RECOMMENDATION SYSTEMS

You have been using this kind of AI for years, probably without noticing. Platforms like Spotify and YouTube rely on AI-powered recommendation engines that suggest songs or videos based on your past behavior. The more you use them, the more accurate they get at predicting your tastes.

- *Use cases:* While you do not interact with recommendation systems directly like you do with a chatbot, they quietly shape your experience by surfacing content you are most likely to enjoy.

Across all these categories, the underlying value is similar: AI helps do things faster, easier, and often more creatively. Whether it is summarizing information, rewriting text, generating fresh ideas, or drafting visuals, AI tools act like assistants that take on the heavy lifting. You do not need to use them all at once. Start with one that fits your needs, and you will quickly see why these tools are becoming everyday companions.

DEBUNKING COMMON AI MYTHS

Whenever there is novelty, myths and misconceptions follow close behind. With AI, the myths can feel especially intimidating because the technology itself seems mysterious. Let's clear the air by focusing on three of the most common fears people have and why they are not quite true.

MYTH 1: AI IS GOING TO TAKE OUR JOBS

This is probably the most common concern. However, while it is true that AI can automate certain tasks, history shows us that new technologies rarely eliminate work. Instead, they change it. When ATMs were introduced, people worried that bank tellers would disappear. Instead, banks opened more branches, and tellers focused on customer service rather than routine transactions.

The same is happening with AI. It can handle repetitive, time-consuming tasks so that humans can focus on higher-value work: strategy, creativity, problem-solving, and personal connection. In this sense, AI will act as an amplifier, giving people more time and energy for the parts of their jobs that require human judgment.

MYTH 2: AI IS ALWAYS RIGHT

Another common misconception is that because AI feels "smart," it must always be correct. But AI is not infallible. It generates answers based on data patterns, not human understanding. This means it can sometimes produce errors, outdated information, or results that sound confident but are flat-out wrong. In fact, this phenomenon is termed "hallucination."

The takeaway is that you should always treat and use AI with caution. While it is extremely good at drafting, suggesting, and inspiring, it still needs oversight. Just as you would not submit an intern's first draft without reviewing it, you should not rely on AI without applying your own judgment. When used thoughtfully, AI is a partner, not a final authority.

MYTH 3: AI IS ONLY FOR TECHIES OR CODERS

This myth stops many people before they even begin. But the truth is that today's AI tools are designed for ordinary users. You do not need to know Python, Java, or any other programming language. If you can write a sentence, you can use a chatbot. If you can describe a picture, you can generate an image.

In fact, one of AI's great breakthroughs is its accessibility. For decades, advanced computing was locked away in research labs or tech companies. Now, powerful AI tools are available to anyone with a smartphone or a laptop. These are no longer niche tools, but everyday tools that are as easy to use as a search engine.

The reality is that AI is not here to replace you but to make your life easier. By taking over the routine and repetitive parts of work, AI frees you to focus on what only humans can do: empathize, create, imagine, and connect. Instead of fearing replacement, think of AI as the assistant who never sleeps, never tires, and never complains about the boring stuff. All you have to do now is learn how to work with it.

CASE STUDIES: EVERYDAY PEOPLE USING AI

Sometimes, the best way to understand a new tool is to see how real people are already using it. Here are three snapshots of how AI is helping ordinary people work smarter and save time:

1. **A teacher using AI for lesson plans:** Maria, a high school history teacher, used to spend hours preparing lesson plans and handouts. With limited time between classes, grading, and extracurricular duties, planning often spilled into her weekends. When she discovered a chatbot tool, she tried feeding it a prompt: *Create a 45-minute lesson plan on the causes of World War I, including discussion questions and a short quiz.* In seconds, she had a draft outline. Maria still reviewed and tweaked the material, but the heavy lifting was done. She also used AI to generate different versions of the quiz for advanced and

struggling students, tailoring the content without tripling her workload. For Maria, AI did not replace her creativity or expertise; it gave her back precious personal time with her loved ones.

2. **A student using AI to understand a difficult topic:** Ethan, a college sophomore, struggled with organic chemistry. Reading the textbook felt like trying to decode a new language. However, instead of giving up, he turned to an AI chatbot and typed, *Explain the concept of chirality in organic chemistry as if I am a beginner.* The chatbot broke it down with simple analogies, comparing molecules to left and right hands. Then Ethan asked follow-up questions, getting increasingly detailed explanations until he finally understood. Later, he used AI to quiz himself with practice questions. The breakthrough was not that the AI gave him new information. What it did was explain the material in a way that matched his learning style. For Ethan, AI became a patient tutor he could turn to anytime, without fear or embarrassment.

3. **A small business owner generating content with AI:** Jasmine runs a small bakery and café. She loves baking but struggles with marketing. Keeping her website updated, writing social media posts, and designing flyers always felt overwhelming. She tried an image generator to create photos of her pastries on colorful backgrounds for Instagram. Then she asked a chatbot: *Write a cheerful Facebook post announcing our new seasonal pumpkin muffins.* In minutes, she had polished, engaging content that would have taken her hours. She still added her personal touch, but the groundwork was already done. Instead of dreading marketing, Jasmine now treats AI as her virtual assistant, helping her connect with customers while she focuses on what she loves: baking.

These everyday examples show the real story of AI. It is not about robots taking over or replacing humans; it is more than that. It is about finding practical ways to save time, reduce stress, and focus on what matters the most.

EXERCISE: IDENTIFY AI IN YOUR LIFE

Before we move any further, let's pause for a quick reflection exercise. One of the biggest surprises for most people is realizing they have already been using AI without even noticing it. Grab a pen and paper and write down five tools or apps you use every day. Don't overthink it. Maybe it's Spotify, Gmail, Instagram, Google Maps, Amazon, TikTok, or Netflix. It could even be the autocorrect on your phone's keyboard.

Now, take a look at your list and ask yourself, *Which of these rely on AI?*

- If you wrote down Spotify or YouTube, those recommendation systems are powered by AI. They study your listening or reviewing history and suggest content tailored to your tastes.

- If you listed Google Maps or Waze, the real-time traffic updates and rerouting come from AI analyzing millions of data points.

- Gmail or Outlook? Their spam filters and smart compose features rely on AI.

- Social media apps like Instagram or TikTok use AI to decide what shows up in your feed.

- Even your phone's predictive text or autocorrect uses ML to anticipate your next word.

Chances are that if not all the apps you use daily have AI, most of them will have it incorporated in some form. You don't need a degree to use them, or even need to know AI was there. It just worked quietly in the background, making your life a little easier. Think about it, how is AI already enhancing your life? Maybe it saves you time by filtering junk emails. Perhaps it entertains you by surfacing the right playlist for your mood. It might reduce stress by helping you navigate traffic.

This simple exercise proves something important: AI is not just some distant, futuristic concept. It is already here and already incorporated into your routine. The difference now is that you are becoming more aware of it, and awareness is the first step toward using it with intention.

JOURNAL PROMPT

Before we wrap up this chapter, let's make this personal. Reading about AI is helpful, but the real value comes when you begin connecting it to your own life. To do that, I would like you to pause for another short exercise.

First, write one sentence about what AI means to you right now. Don't worry about being "right" or sounding smart. This is just for you. Maybe your sentence looks like, *AI feels confusing but exciting.* Or *AI is a tool that might save me time at work.* Or even, *AI makes me nervous because I do not understand it.* Whatever your honest first thought is, put it down.

Later in this book, we will revisit that sentence. You will have the chance to compare how your perspective has shifted after learning, experimenting, and applying AI to your own routines. It is a way of tracking your growth and realizing just how much your confidence with AI can expand.

Next, take a moment to write down a few areas of your life where you would like AI to help you. Think broadly. Some examples to consider include

- At work: Drafting documents, generating ideas, or handling repetitive tasks.

- At home: Organizing schedules, planning meals, or tutoring your kids with homework.

- Personally: Learning a new skill, creating art, or even brainstorming vacation plans.

The goal here is not to make a detailed plan, but to note where the opportunities might be. Once you see those areas written down, AI stops being an abstract concept and starts becoming something tangible, something you could slot into your daily routine in practical and meaningful ways.

This short journal exercise is important because it shifts the focus from fear to possibility. AI is not just "out there" in the news or in tech companies. It is a tool that can fit into your life in selected ways. By naming those areas now, you will be ready to explore how to actually make it work for you in the chapters ahead.

Now that you have reflected on what AI means to you and where it could help, it is time to peek under the hood. In the next chapter, you will understand the mechanics. But don't worry! We are not diving into complicated math or programming. Instead, we will break it down in plain language, with simple analogies, so you can finally understand how AI does what it does. Once you see the moving parts, the mystery fades and the confidence begins.

CHAPTER 2:

HOW AI WORKS (WITHOUT THE JARGON)

AS YOU HAVE SEEN IN THE FIRST CHAPTER, THE GOOD news is that you do not need to be a computer scientist to benefit from AI. You don't need the coding, the technical jargon, or any of the professional knowledge that engineers use. However, here is the thing: If you know just a little about how AI works, you can move from being a passive user to an empowered one.

To make things easier to understand, here is an analogy: Think about your car or motorcycle. Most of us do not know how to rebuild an engine or design a transmission system. But we do know how to drive safely: how to steer, when to brake, and what the warning lights mean. That small amount of knowledge makes us confident drivers. The same is true with AI: You do not need to peek under every hood, but a basic understanding helps you feel in control instead of confused or intimidated.

I remember my own turning point. When I first started using AI tools, I was hesitant. They felt mysterious, like a black box spitting answers. Sometimes, the results were brilliant, and other times, they made no sense. At first, I thought it meant that AI knew something I didn't. But once I learned how these systems were trained, the mystery dissolved. Instead of fear, I felt curiosity. Instead of asking why the answers were wrong, I could see why it made a mistake because of the data it was trained on.

Knowing a little also helps keep you safe. AI, like any other powerful tool, works best when used responsibly. Understanding that AI does not know the truth, but only generates answers based on patterns, teaches you to double-check what it produces. You would not blindly follow GPS directions without glancing at the road signs, and you shouldn't blindly trust AI without applying your judgment.

Most importantly, demystifying AI puts you ahead, where instead of wondering if it will take your job, replace your creativity, or outsmart you, you will see it for what it is: a sophisticated tool that depends on your guidance. That curious, cautious, and confident mindset is the foundation for using AI well.

THE BASICS OF LEARNING: HOW AI IS TRAINED

If you have ever wondered how AI "learns," the answer might surprise you. It does not learn the way people do, with feelings, curiosity, or insight. Instead, it learns by processing enormous amounts of information, called *training data*, and fitting patterns into it.

In this sense, data are the textbooks that they learn with. For a language model like ChatGPT, the training data is made up of huge amounts of text: books, articles, websites, and more. For an image recognition system, the training data is thousands (or even millions) of pictures, each labeled so the AI knows what it is looking at. The more examples an AI sees, the better it becomes at recognizing and generating patterns.

Here is a simple way to think about it: Imagine a child learning to recognize a lion. They would be shown different pictures until they could recognize them. Pictures would include lions from the front, lions lying down, lions in the grass, cartoon lions, and maybe even some stuffed animal lions. Over time, the child learns to recognize the essential features even if every lion looks a little different.

AI works in much the same way. It does not understand what a lion is in the human sense. But after being exposed to enough labeled examples, it can reliably say: "This picture matches the pattern of features I've seen in lions before." This process boils down to three key ideas:

1. **Pattern recognition:** AI is a pattern-spotting master. Just as a child eventually realizes that all lions have whiskers and paws, AI notices repeating structures in the data it's trained on. If it sees the phrase

"peanut butter and..." thousands of times, it learns that "jelly" is likely to follow. In an image, if it sees four legs, fur, and a wagging tail, it learns that the pattern often corresponds to a dog.

2. **Prediction:** Once the AI has recognized patterns, it moves on to prediction. This is where the magic starts to feel real. When you type a question into the chatbot, the system does not know the answer like a human expert. Instead, it predicts what words are most likely to come next based on all the patterns it has seen before. Prediction also explains why AI can be so useful across tasks. Whether it is predicting the next word in a sentence or the movie or song you will enjoy next, the principle is the same: AI takes past patterns and projects them into the future.

3. **Feedback loops:** While pattern recognition and prediction work as an excellent mechanism, alone, they are not enough. The real improvement comes from feedback loops where the understanding is refined. Think back to the child learning about lions. At first, they might point to a tiger and say, "Lion!" You gently correct them, "No, that one has stripes, it's a tiger." With the feedback, the child refines their understanding. AI improves in a similar way, where when it makes wrong guesses, the system adjusts. Over time, corrections help it become more accurate, making it more sophisticated than early chatbots. In some cases, when you use AI, your input becomes part of this improvement, such as when you rate a recommendation as "not helpful" on YouTube or Spotify, the system adjusts to better match your preferences next time.

When you put it all together, training is less like giving AI knowledge and more like coaching it through repetition. The lion analogy is just one example, but it applies everywhere: Show the system enough examples, let it practice predictions, correct its mistakes, and over time, it gets better at the task.

The main point to remember is that AI does not understand the world like we do. It recognizes patterns in data, predicts outcomes, and improves with feedback; that is it. Knowing this demystifies the black box and also empowers you as a user. When you see AI stumble, you will understand why, and instead of losing trust, you can think that this is just a part of how the system learns. Your role is to guide it.

As you understand this simple process, you will already know more than most people about how AI works. This knowledge will allow you to be put in a position where you can use it confidently, safely, and effectively.

HOW LANGUAGE MODELS WORK

One of the most fascinating types of AI, and the one most people interact with daily, is the language model. These are the systems behind tools like ChatGPT, Claude, Gemini, and Microsoft Copilot. They are called *language models* because their specialty is working with text: reading it, analyzing it, and most importantly, generating it. Read on to learn how they work.

TOKEN PREDICTION: THE CORE MECHANISM

Predicting the next word might be a simple idea, but it is quite brilliant if you consider that every language has a structure. When the machine is trained, it is taught to identify these patterns based on what is most common. This means that when you type a question or request into a chatbot, the model does not go hunting through a database to pull out a prewritten answer. Instead, it generates a word-by-word response that predicts what comes next based on what it has learned.

To do this, language models break text down into tokens. These are small chunks of language (sometimes as short as a single character and sometimes as long as a common word or phrase). For example, the sentence, *The dog barked at night*, might be split into tokens like ["The," "dog," "barked," "at," "night"]. The model looks at these tokens and asks itself: *Given the tokens I have already seen, what is the most likely next token?*

It runs a calculation repeatedly, generating one token at a time until it has built a full sentence, paragraph, or even a page of text. This is why language models feel so "fluid." They are not copying and pasting content. Instead, they are predicting, moment by moment, what comes next in sequence.

A GUESSING GAME: FILLING IN THE BLANKS

To make token prediction more comprehensible, imagine you are playing a guessing game with a child where you have to complete each other's sentences. If you say, "Once upon a..." they will probably guess "time." Or if they say "Peanut butter and..." you might guess "jelly." However, if you say, "The birds are..." the word "flying" might come immediately to mind.

What is happening here? You are not pulling these answers from thin air. You are using your experience with language, stories, and culture to predict what word usually comes next. That is exactly what a language model does, only at a very large scale. It has "read" billions of words during training, so when you give it a *prompt*, it can make highly accurate predictions about what should follow.

This happens very commonly when you use a texting app on your phone, for example. Usually, the programs will offer you a few alternatives to continue your sentence after you have entered some information. This is not magic! It is AI and token prediction fully at work.

However, this is also why language models can sometimes produce weird or wrong answers. If you have ever had a friend incorrectly finish your sentence, you know that predictions aren't always perfect. The same happens with AI in different applications, including in your phone texting app: It does not "know" the truth; it just makes predictions based on patterns.

LARGE LANGUAGE MODELS: WHY ARE THEY CALLED "LARGE?"

You have probably heard the term *large language model* (LLM), but have often wondered where the word "large" comes from. In this case, the word refers to both the amount of training and the number of parameters inside the model. Parameters are like the adjustable knobs inside the system that help it recognize and weigh different patterns. The more parameters a model has, the more subtle and complex patterns it can capture.

For example, early language models trained on a few million words had relatively few parameters. They could handle simple tasks but often produced clunky, repetitive text. Modern LLMs are trained on trillions of words, from classic literature to online forums, and can contain hundreds of billions of parameters. That scale allows them to generate responses that feel natural, creative, and context-aware.

To put it another way, imagine learning to write by reading only 10 books. You would get the basics, but your style would be limited. Now, imagine reading an entire library of millions of books, articles, and conversations. Your ability to mimic different voices, tones, and subjects would be much richer. The incredible number of words in the LLMs' "reading list" is what makes them so powerful.

WHY THIS MATTERS FOR YOU

Understanding token prediction and the scale of LLMs gives you a huge advantage as a user. First, it explains why these tools feel so humanlike. Second, it reminds you why they sometimes slip up. This is why it is important to remember that these machines are not thinking; they are predicting at lightning speed. Just as you might finish a friend's sentence with the wrong word, AI can predict incorrectly.

Most importantly, it empowers you to use these tools effectively. The clearer your prompt and the more context you provide, the better the model's predictions. In other words, you guide the guessing game. If you start a sentence with vague or incomplete information, the model has to make a wild guess. This is also why these machines present unreliable results when you use slang or other uncommon languages. The more clearly you set up a sentence, the sharper and more useful these predictions get.

Language models, as of today, are only machines without thoughts and opinions. They are sophisticated pattern recognizers, trained on vast amounts of text, using token prediction to generate new language. Their size, such as in the case of LLMs, makes them flexible and powerful, but the principle is always the same: prediction based on patterns. By seeing it this way, you can strip away the mystery and approach them with confidence. They are incredibly capable machines that work best when you understand the simple game they are playing: predicting what comes next.

IMAGE GENERATORS AND VISUAL AI TOOLS

If language models are impressive because they can generate humanlike text, image generators feel almost magical. You type in a short description, press enter, and within seconds, a brand-new image appears, one that never existed before. It could be a photo, a painting, a cartoon, or even a surreal combination of styles.

So, how does AI turn words into images? The process is different from language models, but the principle is similar: pattern recognition and prediction. Instead of predicting the next word, an image model predicts what pixels should look like based on patterns it has learned from millions of pictures and their descriptions.

During training, the AI sees countless examples of labeled images, such as a picture of a rabbit tagged as "rabbit." It learns that rabbits have fur, four legs, a certain shape, a fluffy tail, and so on. Over time, the system becomes exceptional at recognizing how words connect with visual features. Then, when you type in a prompt like "a whale using a hat," the AI assembles patterns it has learned to create a brand-new image that matches your description.

TOOLS TO EXPLORE

- **DALL·E:** One of the most well-known image generators. It is great for creating anything from realistic photos to quirky illustrations. One of its unique features is "inpainting," which lets you edit parts of an image by describing what you want to change.

- **Canva Magic Studio:** Canva is already a popular design tool, and its AI features make it easy for nondesigners to create professional visuals. You can type a prompt, then drag and drop the AI-generated image straight into a poster, social media post, or presentation.

- **NightCafe:** A creative platform with a variety of styles and community sharing. It is popular for making artistic, dreamlike images and allows you to experiment with different rendering methods and filters.

Each of these tools lowers the barrier to creativity. You do not need to be a trained artist or graphic designer to use them. If you can describe what you can imagine, you can generate it.

TRYING IT YOURSELF

The easiest way to understand image generators is to try them. Here are some simple prompts you can test in tools like those mentioned in the previous section:

- *A watercolor painting of a mountain sunrise.*

- *A cartoon of a car driving a race car.*

- *A professional-looking logo for a coffee shop shaped like a mug.*

- *A futuristic cityscape at night in cyberpunk style.*

Notice that the more specific your description, the more accurate the results. If you just type "whale," you might get a generic picture. But if you say "a whale in the ocean swimming with a hat and a tuxedo," the AI can zero in on your vision.

For many people, image generators are the most tangible way to experience AI's creative power. They are not just for fun (though they are certainly entertaining). Businesses used them for marketing visuals, teachers for classroom materials, writers for book covers, and entrepreneurs for product mockups.

What is revolutionary is the accessibility. In the past, creating a custom illustration meant hiring an artist, paying significant fees, and waiting days or weeks. Now, anyone can create something unique in seconds. AI does not replace artists (just as cameras did not replace painters), but it expands what is possible for everyone. Professionals can use it to prototype ideas faster, while beginners can finally bring their visions to life.

At the same time, though these tools are quite impressive, you must receive an important reality check: AI tools can also be wrong, strange, and even misleading. Understanding why helps you use them confidently, safely, and correctly.

WHY CAN AI BE WRONG (AND SOMETIMES WEIRD)

So far, AI may sound like a miracle machine: writing essays, generating art, and even composing music. However, the truth is that AI is not perfect—

far from it. Sometimes, it is not even close. To understand why, you need to remember one simple principle: AI is only as good as the data it is trained on.

During training, AI systems are exposed to massive amounts of information, as you have already seen. These come from books, articles, images, videos, and many other online information. However, they do not "understand" the content in a human sense; all they do is look for patterns. If the training data contains errors, biases, or gaps, the AI will reflect these flaws. Just like a student who learns from a flawed textbook, an AI trained on imperfect data will make imperfect predictions.

AI's key limitation is that it does not know truths. It doesn't "understand" the world the way we do. It does not have experiences, values, or judgment. Instead, it produces answers by predicting what *sounds right* based on the patterns it has seen before. That distinction matters because if you ask an AI chatbot about a scientific fact, it might generate a convincing response, complete with citations. However, these citations might not exist because the AI can make them up.

Why? Because the model has learned that facts are often followed by references, it generates something that looks like a reference, whether it is real or not. In other words, AI can be confidently wrong, and, when it is, it can fool even savvy users. Some of the most memorable examples of AI getting it wrong have gone viral:

- **The Pope in a puffy jacket:** In 2023, an AI-generated image of Pope Francis wearing a stylish white puffer jacket spread across social media. Many people thought it was real until it was revealed to be a creation of an image generator. The AI knew how to combine patterns of "Pope," "white clothing," and "streetwear style," but it had no sense of appropriateness or truth.

- **Invented quotes:** Language models tend to invent quotations. Ask them to provide what a historical figure said about a specific topic, and they might produce a well-written but completely fabricated line. The AI does not want to deceive but is simply generating text that fits the pattern of "famous person + quote," whether or not the quote exists.

- **Odd visual details:** Image generators sometimes produce surreal results: people with too many fingers, text that looks like gibberish, or impossible physics (such as a shadow pointing the wrong way). These mistakes happen because the AI is stitching together patterns of pixels, not reasoning on how the physical world works.

These examples can be funny, but they highlight a serious point: AI is not based on reality. Without human oversight, it can spread misinformation or create misleading content.

WHY FACT-CHECKING IS IMPORTANT

Because AI has no built-in sense of truth, the responsibility falls on the users to verify its output. AI is the talented but unreliable intern that can produce brilliant drafts and creative ideas, but it needs supervision. Here are a few practical ways to stay in control:

- **Double-check facts:** If AI gives you a statistic, quote, or reference, verify it using a trusted source.

- **Treat images with caution:** Just because a picture looks real, it does not mean it is. If something feels "too perfect," it might be AI-generated.

- **Look for consistency:** AI sometimes contradicts itself in longer responses. Reading carefully can reveal red flags.

- **Use your expertise:** If you know the subject well, trust your instincts. If something feels off, it probably is.

While it may feel unsettling that AI can be wrong, in many ways, it is good news. If AI were perfect, it would have replaced more human roles outright. Its imperfections remind us that these systems are tools designed to assist, not dictate.

As you become better able to determine where AI goes wrong and why, you are already ahead of the curve. Instead of being dazzled or deceived, you can approach AI with healthy skepticism. Use it for what it does best: generating drafts, sparking ideas, and making predictions, and remember that you, the human, are the final editor and decision-maker.

ACTIVITY: TEACH AI TO EXPLAIN A CONCEPT

One of the most powerful features of LLMs is their ability to adapt explanations to different audiences. Unlike a static textbook, AI can adjust both *what* it says and *how* it says it. This makes it a fantastic learning partner, regardless of whether you are trying to understand something new yourself or communicate it clearly to others. In this activity, you will use a chatbot to explain the same concept at three different levels: for a child, for a peer, and for an expert:

- **Step 1: Pick a complex topic:** Choose any topic that feels challenging but interesting. It could be something from your work, your studies, or even a concept you always wanted to grasp better. For our example, let's pick photosynthesis.

- **Step 2: Ask AI to explain it to a child:** Your first prompt could be, *Explain photosynthesis to a 7-year-old.* The AI might respond with something like: "Plants are like little chefs. They take sunlight, water, and air, and mix them together to make food for themselves. The green part of the plant acts like a kitchen where this happens." Notice the simplicity of the explanation that has short sentences, playful analogies, and no jargon. This level is great when you want to break down intimidating ideas into something you can actually visualize.

- **Step 3: Adjust for a peer-level explanation:** Now, adjust the prompt: *Explain photosynthesis as if you are talking to a high school student.* The AI might say, "Photosynthesis is the process plants use to make their own food. They take in carbon dioxide from the air and water from the soil. Using energy from sunlight, they produce glucose, which is a type of sugar, and release oxygen as a by-product." The AI keeps the explanation accurate but still approachable. It uses proper scientific terms, but without going too deep into the molecular details.

- **Step 4: Ask for an expert-level explanation:** Finally, challenge the model: *Explain photosynthesis to a graduate student in biology.* Now, the answer may include advanced terms like chloroplasts, thylakoid membranes, electron transport chains, and ATP synthesis. It might discuss light-dependent versus light-independent reactions, or the role of specific enzymes. At this level, the AI assumes the reader has a strong foundation and is ready for precise, technical language.

UNDERSTANDING THE EXERCISE

By running the same topic through these three levels, you can see something fascinating: AI not just recalling facts; it is also tailoring your communication style. I can shift tone, vocabulary, and depth depending on the audience you request. This adaptability is one of the biggest reasons language models feel so useful. Whether you are a student struggling to understand a difficult chapter, a teacher preparing different lesson levels, or a professional trying to simplify complex information for a client, AI can be flexible to meet the moment.

TRY IT YOURSELF!

Now, it's your turn to pick a topic you find tricky. These could range from how mortgages work to what blockchain is or the basics of climate change. Run it through the same three steps: child, peer, and expert. Compare the answers. You may notice new ways of seeing the same idea. Sometimes, the child-level explanation gives you the clarity you did not know you needed.

As you will see, AI is not just about facts, but also about communication. It can take the same idea and shape it to fit whoever is listening. That makes it a tool for understanding and a bridge for sharing knowledge across levels of expertise. The better you get at prompting AI to adjust its tone and style, the more you will see its value as more than a static encyclopedia.

JOURNAL PROMPT

Now that you have seen how AI works, it is time to pause and reflect. Grab a notebook or open your notes app. Write down your honest response to this question: *Now that I understand how AI "thinks," how will it change the way I use it?* There is no right or wrong answer here; what matters is that you anchor this new knowledge in your own life.

You have moved from seeing AI as a black box to understanding it as a tool that predicts, creates, and occasionally makes mistakes. That awareness gives you the power to use AI with confidence instead of hesitation.

Next, jot down one or two areas where you would like to experiment with AI differently now that you understand its mechanics. These small steps can build into meaningful changes in how you work, learn, and create. This is all possible because you now know enough to demystify its behavior, laugh at its quirks, and use it wisely. But this does not mean that there is no fear or overwhelm. Read on to explore these feelings and how to overcome them.

CHAPTER 3:

OVERCOMING FEAR AND SKEPTICISM

YOU ARE NOT ALONE IF YOU LOOK AT AI AND FEEL UNEASY or skeptical. For many people, the word "AI" stirs up more anxiety than excitement. Instead of imagining tools that help us, the first images that come to mind are robots taking jobs, algorithms making secret decisions, or a future that feels out of our control. It is less about the technology itself and more about the emotions it triggers: uncertainty, overwhelm, and sometimes even fear for our livelihood.

This is not unusual. History shows that every major innovative leap is accompanied by a wave of doubt before it brings benefits. When electricity first entered homes, families were terrified it would spark fires or cause harm. When personal computers arrived in offices, employees whispered about whether machines would replace them. Even the internet, which today feels as natural as breathing, once carried with it the same mix of fascination and fear. AI is simply the newest chapter in the book of history.

I have witnessed some of these emotions firsthand. Maria, for example, was a university teacher who was convinced that AI would undermine her role in the classroom. When she discovered her students using chatbots to write essays, she felt a flash of panic: *If machines can do their work, what's left for me to teach?* For weeks, she avoided the subject entirely, hoping it was just a passing trend. Later, almost by accident, she tried AI to draft a lesson plan. To her surprise, the tool didn't erase her role, but *helped her* get a head start, freeing her to focus on the creative and human sides of teaching that no chatbot could replace.

James, a healthcare worker in his fifties, had a different kind of resistance. He had grown comfortable with the systems he already knew and dismissed new AI tools as something for younger, tech-savvy colleagues. However, when his hospital introduced an AI assistant for paperwork, he gave it a cautious try. The relief was immediate: what used to take hours of form-filling was suddenly done in minutes, leaving him more time with his patients. Far from replacing him, AI gave him the part of the job he loved the most.

Then there is Sandra, a grandmother who once thought AI belonged only in science fiction movies. When her daughter showed her how an AI assistant could quickly generate shopping lists and meal plans, she laughed it off. But after seeing how much stress it saved her week after week, she admitted that the technology made her life easier in an accessible manner.

These stories reflect what so many of us feel: a mix of skepticism, caution, and maybe even dread when we first hear the term "AI." But fear does not have to be the end of the story. More often than not, it is just the beginning. Once people try AI for themselves and on their own terms, for problems that matter in their daily lives, the mystery starts to dissolve. The black box becomes a tool, and AI becomes your ally.

That is what this chapter is about. We will unpack the most common AI-related fears, look at their origin, and most importantly, explore how to move past them. The truth is simple: The best way to overcome fear of AI is not to ignore or debate it endlessly, but rather to use it and see for yourself that it is not here to take over your life.

TOP 5 AI FEARS

Fear is not something to brush off when it comes to AI. It is real, it is valid, and it often comes from stories we have heard or headlines we have read. The good news is that most of these fears, while understandable, are not the full picture. Here are five of the most common worries people share about AI and what their actual nature is.

1. *"I'll be replaced"*: This is probably the number one fear, and it makes sense. Every time there is a giant technological leap, people wonder if they will become obsolete. We saw this with the Industrial Revolution, with computers, and the internet. The truth is that AI

will change the way work is done, but in most cases, it does not replace people. What it does instead is to help people shift their focus, especially by taking over repetitive and routine tasks. When this is implemented, you can make your work meaningful by using human judgment, empathy, and creativity. Rather than seeing AI as a rival, think of it as a power tool that makes your job faster and less tedious.

2. *"I don't understand it, so I can't use it"*: This fear is rooted in overwhelm, starting with all the tech jargon. Algorithms, neural networks, ML, and others can feel intimidating, especially if you do not come from a technological background. But here is the thing: You don't need to understand AI to use it efficiently; after all, you do it all the time. You might not understand cars, but you can drive; you do not know the engineering behind touch screens, but you can use a phone; you cannot build a computer, but you can use it at home or work. The same applies to AI tools, where a little basic knowledge helps, but you do not need a degree to use them. In fact, most AI tools today are quite simple to use: You type in what you want, press enter, and they do the rest.

3. *"It's stealing people's work"*: This concern surfaces especially with AI-generated art, writing, music, and other creative areas. The worry is that AI is copying human creativity without permission. However, while real debates are happening about ethics, copyright, and regulation, it is important to understand how AI works. AI does not "steal" work in the sense of copy-pasting. Instead, it learns patterns and creates something new from that. That does not mean that ethical questions do not exist; they do and are being debated right now in governments, courts, and industries. However, on a personal level, using AI for brainstorming, drafting, or designing does not make you a thief. You are using a tool that, like all tools, is built on collective knowledge.

4. *"It knows too much about me"*: Privacy is a very real concern in our digital age, and AI can feel like one more step toward a loss of control. After all, if it is so good at predicting things, does it mean it's reading your mind? Not exactly. Most AI tools do not know you

personally. They work from the general patterns in their training data. When you ask a chatbot a question, it's not pulling information on your life; it's predicting an answer based on the billions of words it has already seen. A good rule of thumb is to treat AI tools like any online service: Don't share sensitive information you wouldn't want stored, and keep your personal details out of any queries.

5. *"It's only for tech people"*: Perhaps the most limiting fear of all is the idea that AI belongs to experts: coders, engineers, and the Silicon Valley crowd. The reality could not be further from the truth. AI tools are being built with everyday people in mind. Teachers are using them to correct and create student homework, and parents are using them to organize their busy schedules. None of these people need to be "techies," they only need curiosity and a willingness to try. Always keep in mind that AI is not about your job title or your age. Not at all. It is about finding ways to make your life smoother.

All these fears come from a very human place, and each one has a more balanced story beneath it. While fear thrives in the unknown, when you peek behind the curtains, you can see AI for what it is: not a threat waiting to take over, but a tool waiting for you to put it to work.

MINDSET SHIFT: AI AS A POWER TOOL, NOT A THREAT

One of the most helpful ways to reframe AI is to stop seeing it as a rival and start seeing it as a tool, a formidable one, in fact. Imagine when electricity first spread into homes and businesses. At the time, people worried it would replace workers or even be dangerous to human life. Instead, electricity became the backbone of new industries: lighting, refrigeration, communication, and manufacturing. Instead of replacing people, electricity empowered people and shifted them toward new ideas.

AI is on a similar path where it does not eliminate the need for human qualities. What it does instead is take on the heavy lifting or repetitive, time-consuming tasks so we can spend more time doing what we are uniquely good at. A writer can spend less time stuck on first drafts and more time polishing ideas. A doctor can spend less time on paperwork and more time with patients. A teacher can spend more time actually teaching.

This is where the idea of "human + AI = augmented intelligence" comes in. Instead of AI meaning machines replacing us, it should be about machines supporting us. Augmented intelligence means AI fills in the gaps, handles the grunt work, and offers new perspectives while bringing context, empathy, and critical thinking. It is not about who is smarter and who does what best; it's about what can be achieved together.

When you shift your mindset in this way, AI is no longer a looming threat but is embraced as a promising opportunity. You do not need to compete with it if you collaborate with it. Just as electricity lit up the world, AI can illuminate new possibilities in your own life and work.

CASE STUDIES: HOW SKEPTICS BECAME CONFIDENT USERS

Sometimes, the best way to understand how a mindset shifts is through stories. AI can sound abstract until you see how ordinary people who once resisted using the technology learned to make it a part of their daily routines. Here are three examples of skeptics who discovered, in very different ways, that AI was not here to take over what they do.

THE NONPROFIT DIRECTOR WHO HATED "BUZZWORDS"

Marcus had been running a small nonprofit for over a decade. His focus was on community programs, not technology. Every time someone mentioned AI at a board meeting, he rolled his eyes. "We do not need flashy buzzwords; we need results," he used to say. One afternoon, while preparing a grant application, Marcus hit a wall. The proposal was due in 48 hours, and he was staring at a blank page.

A colleague suggested he try a chatbot to draft an outline, and he reluctantly gave it a go. To his surprise, the tool produced a clear, professional draft in minutes. It was not perfect, but it gave him a solid foundation. Marcus polished it with his own knowledge and voice, and for the first time in years, he submitted a grant proposal ahead of schedule.

THE RESTAURANT OWNER WHO WAS "TOO BUSY"

Lina owned a family restaurant and felt stretched thin managing menus, staff schedules, and social media. When her nephew suggested she use AI to help with marketing, she laughed it off and said she was too busy to learn how new gadgets work. But when sales dipped in a slow season, she gave in. She asked an AI tool to generate a week of Instagram captions and menu highlights. Within half an hour, she had polished posts ready to go, which would usually take her an entire evening.

The results surprised her: Her engagement improved, and she also felt less stressed. Lina now uses AI to brainstorm seasonal specials, draft emails to loyal customers, and even create recipe cards for staff training. By the end of the season, she had more control of her agenda and other restaurant business, allowing her to work on customer service and other tasks that required her input.

THE RETIRED ENGINEER WHO MISSED TEACHING

After retiring, Paul missed mentoring younger colleagues. He still loved learning, but he found modern technology overwhelming. When a friend showed him how AI could explain topics at different levels (child, peer, and expert), he was skeptical.

Out of curiosity, he asked a chatbot to explain quantum mechanics in simple language. The answer was clear, playful, and surprisingly accurate. He then asked for an "expert" opinion and was impressed by the depth. Soon, Paul started using AI to refresh his own knowledge and even built short learning guides for local students. He realized that AI was not replacing teachers, but actually extends what teachers do, facilitating a way to share knowledge.

What links Marcus, Lina, and Paul is not their backgrounds or jobs, but that each started to use AI with skepticism. However, once they used to solve a problem they actually cared about, the fear softened. The shift came from experience, which is the key to overcoming fear. AI is not about memorizing definitions or reading manuals. Once you have had that first win, the fear loses its grip.

ACTIVITY: SOLVE ONE ANNOYING PROBLEM WITH AI

The fastest way to overcome fear of AI is not to just read about it; you also have to put it to work in something small but useful. Do not start with a life-changing project or a job-defining task. Use it for one of those little annoyances that eat up your time and energy.

By the end of this exercise, you will see what it feels like to let AI take the load off your shoulders. Here are three options to choose from. Pick one that feels the most relevant to your life and give it a try.

OPTION 1: ORGANIZE MY WEEK

If you have ever looked at your calendar and felt your head spin, this one is for you. Here is a prompt you can try:

I have a busy week ahead. Here are my commitments (list them here with days and times). Can you create a balanced schedule that leaves time for meals, exercise, and at least 30 minutes of relaxation each evening?

The AI might come back with a color-coded outline or a simple daily schedule that spaces things out. You will probably tweak it, but the heavy lifting of structuring your time is already done. Instead of staring at a blank calendar, you start with a clear plan. Notice how much mental energy you saved by offloading the initial organization. Even if you adjust the details, the hardest part is done.

OPTION 2: WRITE A POLITE EMAIL TO RESCHEDULE

Few things feel more awkward than having to cancel or reschedule plans. Whether it is moving a meeting or postponing a coffee date, we often agonize over the wording. One of the prompts that you can try is the following:

I need to reschedule a meeting with my colleague on Friday. Please write a polite and professional email that apologizes for the inconvenience, suggests two alternative times, and keeps a warm tone.

The AI will generate a draft that strikes the balance for you. From there, you can add your own touches, such as a personal note or a different time slot, but the awkwardness of starting from scratch disappears. Once this is done, think about how much time you have spent in the past overthinking simple emails. Here, AI turns a 20-minute chore into a 2-minute edit.

OPTION 3: SUGGEST 5 DINNER IDEAS UNDER 30 MINUTES

End-of-the-day fatigue is real. After work, errands, and family responsibilities, the last thing most of us want to do is figure out what is for dinner. You can try this prompt to help you out:

Give me 5 healthy dinner ideas that I can make in under 30 minutes. Please include a list of ingredients and simple instructions for each.

The AI might suggest quick stir-fries, pasta dishes, or sheet-pan meals, all with the steps laid out for you. You can even refine the request and ask to make it vegetarian or for the machine to use the ingredients you have at hand. This will make that dreaded nightly decision leave your plate (pun intended), and instead of scrolling through recipes or ordering takeout, you have practical options in seconds.

WHAT YOU LEARNED

If you tried one of these exercises, you probably noticed two things: the relief of outsourcing the hard part and the time saved. The first was that you did not have to build from zero, since you had a head start. The second was that what once took half an hour or more has now been trimmed down to minutes. These are small "wins," but they matter.

Every time AI saves you from frustration on a little task, you build confidence. As that confidence grows, you will begin to see bigger opportunities. Today, it is a dinner plan; tomorrow, it can be drafting a proposal, brainstorming a lesson plan, or summarizing a report. The important part is that you have moved from fearing AI as something abstract and overwhelming to using it as a practical ally.

Each time you let AI solve an annoyance, you chip away at the fear and replace it with trust. AI does not have to be about the future of work or massive societal change. Sometimes, it is just about freeing up your evening so you can relax with your family instead of wrestling with your inbox or calendar, and this is where the real magic begins.

BUILDING AI CONFIDENCE ONE STEP AT A TIME

Confidence does not appear overnight. Think back to the first time you did something new. The chances are that you were not an expert right away. You probably felt nervous, perhaps even clumsy. However, the more you practice, the more the fear is erased, and before long, the thing that once felt intimidating becomes second nature. Learning to use AI is no different. The good news is that you do not need to "master" AI in one go. In fact, the best way to build confidence is through small, steady steps that let you experience wins without pressure. Here is how to start.

CREATE A SAFE ENVIRONMENT

First, give yourself the chance to experiment. Think of your early interactions with AI as practice sessions, not a high-stakes exam. You do not have to use it for work right away or apply it to something critical. Start in a safe place: plan a weekend, brainstorm creative ideas, or ask for suggestions. When the stakes are low, the fear of "messing up" evaporates. You can laugh at the weird or wrong answers, shrug off the occasional nonsense, and treat the experience as a learning curve. Over time, you will notice that those small, harmless experiments build a foundation of familiarity that makes tackling bigger tasks feel less daunting.

WHAT TO IGNORE, WHAT TO BE CURIOUS ABOUT

Part of building confidence is knowing what not to worry about. It is easy to get overwhelmed by headlines about AI ethics, technical jargon, or heated debates about the future of work. While those conversations are important, they can also distract you when you are just starting out.

For now, you can safely ignore the jargon, the fear-mongering, and the pressure to be perfect. You do not need to know how a neural network works to use AI, just like you do not need to know how the TV works to watch it. Articles claiming that AI will replace everyone make great headlines, but they still do not reflect the everyday reality of how people actually use it. Finally, if your first prompts do not work, that is normal. AI is a dialogue, and you will learn how to phrase things better over time.

Instead, focus your curiosity where it counts:

- prompts and responses

- use cases in your life

- patterns in mistakes

Curiosity is your biggest ally. Instead of worrying about what you do not know, lean into exploring what is possible.

EMBRACE MISTAKES AS PART OF THE PROCESS

One of the fastest ways to lose confidence is by expecting perfection. The reality is that AI will sometimes get it wrong. It might misunderstand your request, make up a fact, or produce something that feels off. Instead of seeing these mistakes as failures, treat them as part of a journey. Remember: AI is a prediction machine, not a truth machine. When it makes a mistake, it is not "broken." What it is doing is offering you a chance to refine your input. Often, the fix is as simple as clarifying your prompt or asking to try again with more detail.

Mistakes also highlight where your human strengths shine. You are the editor, the fact-checker, the one with judgment and context. AI provides the draft or the spark, but you shape it into something valuable. It is similar to when you are learning a new language: At first, you stumble over grammar and pronunciation, but these mistakes are not wasted. What is happening is that you are practicing to reach fluency. With AI, each quirky response is an opportunity to learn how to communicate with it more effectively.

STEP INTO CONFIDENCE

Here is the real secret: Confidence grows with use. The more you let AI handle small annoyances or routine tasks, the more natural it feels. Over time, you stop thinking of it as a mysterious technology and start seeing it just as another everyday tool. So, start small: Ask it to draft a shopping list, organize your day, or summarize a long article. Celebrate the minutes saved and the frustrations avoided. With each small win, you are training yourself to trust the process.

What happens if something does not go as planned? That is not a failure, but progress. Every experiment, good or bad, brings you closer to fluency. Before long, you will find yourself not only using AI with confidence but also teaching others how to do the same. Confidence, remember, is not about knowing everything, but being willing to try, learn, and adapt. With AI, those little steps are all it takes to transform fear into empowerment.

JOURNAL PROMPT

Take a pause here. You have read about the fears, unpacked the myths, and even tried your hand at using AI for small but meaningful wins. This is the perfect time to reflect and anchor your progress. Grab a notebook or your phone and write down the answer to this first question:

What small win did I achieve by using AI today?

The size of the win does not matter. What matters is that you noticed it and that you allowed yourself to try. Naming that win makes it real, and it is proof that you have already moved from fear to action.

Now for the second question:

How would I explain AI to someone more skeptical than me?

This is powerful because teaching (or imagining yourself teaching) is one of the best ways to deepen your own understanding. Picture a friend, a co-worker, or even a family member who still sees AI as a threat. How would you put it into words? Framing it for someone else helps you clarify what AI means to you right now, and it also reinforces the truth: You have moved from hesitation to confidence, and you are already capable of guiding others through the same journey.

As we close this chapter and the first part of this book, look back and review the important steps you have taken. You faced your fears, saw them for what they are, and replaced them with curiosity, small wins, and growing confidence. You no longer see AI as something distant and intimidating; you have started to see it as something personal, practical, and even helpful.

This is where Part II begins. You will learn how to use AI not just as a novelty or a helper for chores, but as an assistant that works alongside you, organizing your tasks, supporting your creativity, and amplifying what you already do best. Are you ready to start really diving into the world of using AI? If so, go ahead to the next part!

PART II:

PRACTICAL AI FOR PERSONAL PRODUCTIVITY

CHAPTER 4:

YOUR PERSONAL
AI ASSISTANT

BE HONEST: HOW OFTEN DO YOU FEEL STRESSED BECAUSE you have so much to do and your time is stretched thin? Between work, family, errands, and the endless stream of emails and messages, there are never enough hours in the day. You wake up with the best intentions, but by the time evening rolls around, the to-do list is still half full and your energy is gone.

Sarah is a project manager and mom of two. Her mornings were a blur of school lunches, last-minute work calls, and unanswered emails. By the end of the week, she often felt like she was treading water: busy all the time but never quite caught up. When a colleague suggested she try an AI assistant to organize her schedule and draft routine emails, she shrugged it off, saying she was managing just fine. However, when she tried it, Sarah saw AI gave her something she was missing: breathing room. Tasks that used to take hours melted down to minutes, and for the first time in months, she left the office on a Friday with a clear desk and a clear head.

The secret here is knowing that Sarah's story is not unique. We all have some sort of assistant, from setting timers in Siri, adding items to Alexa's shopping list, or using Google or social media to suggest quick replies to messages. The difference is that now AI has become smarter, easier, and more accessible to users. Instead of just reminding you to buy more milk or proofreading an essay, these tools can now help with business plans, with planning for your week, or optimizing your exercise schedule. The best part is that they do not eat, complain, sleep, or ask for a raise.

As I've reiterated, you definitely do not need to be techy to use them. If you can type a question, you can use an AI assistant. This chapter shows you exactly how to do that, step-by-step, with simple tools that are free or nearly free. You will learn how to offload the cognitive clutter that is time-consuming: writing, sorting, scheduling, summarizing, and preparing. By the end, you will learn how to turn AI into a reliable helper who saves you hours every week.

Therefore, if your days often feel like Sarah's, with too much to do and not enough time, this chapter is your turning point. You will now have access to a cost-free assistant that is ready to take a load off your shoulders at any time. All you have to do is start asking.

CHOOSING THE RIGHT AI ASSISTANTS (FREE OR FREEMIUM OPTIONS)

When people hear "AI assistant," they often imagine something complicated or expensive. The reality is much simpler and much more exciting. Right now, some of the most powerful AI assistants are available for free, or nearly free, to anyone with an internet connection. These tools are not locked behind advanced coding skills or pricey subscriptions. If you can open a web browser and type a sentence, you are already equipped to start.

This section will introduce you to four of the most reliable, widely used AI assistants: ChatGPT, Perplexity, Claude, and Google Gemini. Each one has its own strengths, quirks, and best cases. Think of them like team assistants, each specializing in a slightly different task. By the end, you will know how to set them up, try them out, and organize them so they are always just a click away.

CHATGPT

OpenAI's ChatGPT is one of the most popular LLMs in the world. It works like a conversation partner: You type in a question, request, or prompt, and it responds in a natural, humanlike language. It is best for general problem-solving, writing drafts, brainstorming, summarizing text, and practicing explanations at different levels. It is the "all-rounder" of AI assistants, great for both personal and professional tasks. Here is the step-by-step process to set it up:

1. Go to chatgpt.com in your web browser.

2. Sign up for free with your email, phone, or Google, Apple, or Microsoft account (if you want access to your conversation history).

3. Once logged in, you'll see a simple text box. That's your workspace. Just type something like, *Help me plan a 3-day trip to New York City,* and hit Enter.

4. You're now chatting with your assistant.

The chatbot (at the time of writing) uses GPT-5 by default, which switches to the mini (but still highly capable) version for free users once usage limits are reached. If you ever want more advanced features, a paid option is available, but unnecessary to start.

PERPLEXITY

Perplexity is like an AI-powered search engine. While ChatGPT excels at conversations and creativity, Perplexity shines at finding real, current information. It does not just give you an answer—it shows you where the information comes from, linking to sources you can verify. It is best used for research, quick fact-checking, current events, and any situation where you want trustworthy, sourced answers. To set it up, follow these steps:

1. Go to perplexity.ai in your web browser.

2. No account is required to get started, but you can sign up for free to save your history.

3. Type your question into the search box, just as you would on Google: *What are the current best practices for hybrid work policies?*

4. Perplexity will generate an answer and provide citations from reliable sources at the bottom.

This ability to show its work is what makes Perplexity stand out. If you are ever worried about whether AI is making things up, this is the assistant to use.

CLAUDE

Anthropic's Claude is another conversational AI similar to ChatGPT. What makes it special is its focus on safety, friendliness, and handling large amounts of text. You can paste in long documents such as essays, articles, and even full transcripts, and Claude will summarize or analyze them with ease. It is best used for summarizing large chunks of text, drafting professional communication, and brainstorming ideas in a friendly, approachable style. Follow these steps to start using it.

1. Go to claude.ai in your web browser.

2. Sign up with your Google account or email address (it's free).

3. Once inside, you'll see a chat box similar to ChatGPT. Try pasting a long article and typing: *Summarize this in 3 bullet points.*

4. Claude will give you a digestible version of the text in seconds.

Claude is especially used by students, researchers, or professionals who regularly deal with long documents. Instead of slogging through dozens of pages, you can let Claude handle the heavy lifting and then focus on the insights.

GOOGLE GEMINI

This AI tool, formerly known as Bard, is directly integrated into the Google ecosystem. That means it is connected to tools many people already use daily, such as Gmail, Google Docs, or Google Drive. It's conversational like ChatGPT, but also has the advantage of tapping into live web data. It best serves those who already use Google tools, as well as for quick brainstorming, drafting emails, and using AI to work with their existing files and documents. Setting it up is simple by following these steps:

1. Go to gemini.google.com in your web browser.

2. Sign in with your Google account (most people already have one) for conversation history tracking, or start chatting right away without an account.

3. Start typing your requests into the chat box. For example: *"Summarize the last 5 emails in my inbox and draft a reply."*

This makes it incredibly practical for everyday communication and productivity. If your work already lies in Google Drive, Gemini can feel like a natural extension of your workflow.

By now, you have met four different AI assistants, each with its unique abilities:

- **ChatGPT:** The versatile all-rounder for writing, brainstorming, and problem-solving.

- **Perplexity:** The fact-checker and researcher that always shows its sources.

- **Claude:** The friendly summarizer that handles long text with ease.

- **Google Gemini:** The Google-native assistant that blends seamlessly with email, the internet, and docs.

You do not have to pick just one. In fact, most people find it useful to have two or three bookmarked and ready to go, using an appropriate one depending on the task at hand. It's like having a team of assistants in your pocket, each prepared to solve a different problem. To make these tools part of your daily routine, create a dedicated folder in your web browser's bookmarks. Call it "AI Assistants" and inside, add links to these four tools. That way, whenever you face a task, you are only two clicks away from help. The easier they are to access, the more likely you are to use them regularly.

EMAIL AND COMMUNICATION MANAGEMENT

If there is one task that seems to consume more time than it should, it is managing email. Personal or professional, our inboxes often feel like bottomless pits: endless threads, vague subject lines, and the dreaded wall of text emails that require 20 minutes of concentration just to understand. Add to that the pressure of sounding polite and professional in every reply, and it is no wonder emails are one of the greatest productivity drains of modern life. Studies reveal that in a business setting, "70% [of employees] believe it's one of the biggest productivity drains in the workforce and a further 73% say too much time is spent trying to find emails" (Just In: Email, 2021).

This is exactly where AI can step in: not to replace your judgment, but to save you the mental energy of drafting, summarizing, and formatting. By handling the "heavy lifting" of communication, AI frees you up to focus on what really matters in the message. Generally speaking, AI can simplify communication in two main ways:

- **Drafting responses quickly:** By providing the tool with intent, you can avoid agonizing over wording and create a professional draft you can tweak.

- **Summarizing long threads:** Instead of scrolling through endless back-and-forth, you paste the conversation into an AI tool and ask for the "too long; didn't read" (TL;DR) version.

Both of these functions sound small, but in practice can save you hours each week and reduce decision fatigue.

USE CASE 1: WRITE EMAILS FASTER

One of the most common frustrations is staring at a blank screen, trying to compose a polite reply. AI takes away that pressure by giving you a starting point. You still add your personal voice, but you are no longer starting from zero. Here are a few prompts you can try in your preferred AI tool:

- *Write a polite email to my coworker letting them know I cannot attend Friday's meeting and suggesting 2 alternative times.*
- *Draft a professional but warm reply to a client who asked about the status of a project. The update is that we are on track and will deliver by next Wednesday.*
- *Write a concise thank-you note after a job interview. Keep it friendly, express gratitude, and highlight my enthusiasm for this role.*

Within seconds, AI will give you a draft that you can quickly edit to match your tone. What used to take 20 minutes of second-guessing now takes 2 minutes of editing.

USE CASE 2: SUMMARIZE LONG EMAILS OR THREADS

Another common frustration is wading through long emails or group threads where everyone has chimed in. AI can strip away the clutter and give you the gist. Look at these prompts and try them out for yourself:

- Copy the email text (removing names and sensitive info) and paste it into your AI assistant. Then type: *Summarize this email in three bullet points and tell me what action is required from me, if any.*

- If it is a thread, you can say: *Summarize this thread of emails into the key decisions made, outstanding questions, and next steps.*

The relief here is enormous. Instead of rereading the chain, you get clarity in seconds.

Privacy Reminder

Before you paste any information into an AI tool, there's one important step: protect sensitive information. AI tools are improving their privacy safeguards, but as a best practice, remove names, financial details, or confidential content. Replace them with placeholders if needed (e.g., "Client A," "Project X"). This keeps your information safe while still letting the AI process the general content.

Email is where many people first tangibly feel AI's power. You do not have to be creative, technical, or even adventurous to benefit. All you are doing is letting the assistant handle the boring parts like the drafting and the summarizing, so you can get back to the human part: connecting, deciding, and moving things forward.

SUMMARIZING, SCHEDULING, AND TASK LISTS

One of the hidden costs of modern life is not just work itself, but all the mental load around the work: reading, sorting, planning, and organizing.

Emails pile up, articles wait to be read, reports gather dust, and task lists sprawl across sticky notes and apps. For many people, the problem is not that they cannot do the work, but that they are drowning in prework. However, when you throw AI into the picture, the situation changes.

AI can be a lifesaver with the right prompts, especially to take on the cognitive clutter, distilling long documents into key points, structuring your week into something manageable, and breaking large projects into small and doable tasks. In other words, it helps you think more clearly by doing some of the thinking for you. Let's walk through three of its most powerful uses to evaluate how it can help you.

SUMMARIZING PDFS, ARTICLES, AND REPORTS

If you have ever opened a report or research paper, skimmed the first page, and thought, *I will get back to this later*, but never returned, you are not alone. Long documents can feel intimidating, especially when you just need the highlights. However, with AI, this all changes.

All you need to do is copy and paste a test from an article, PDF, or report into your assistant and ask it to summarize. Many tools like Claude and ChatGPT can even handle file uploads, letting you drop in entire documents. Imagine you are given a 30-page market research report for work. Instead of spending hours reading line by line, you paste it into an AI assistant and ask: *Summarize this report in 5 bullet points, highlighting the main findings and recommended actions.*

Within seconds, you will get a concise overview and even refine it further using commands such as

- Now rewrite this summary for a nontechnical audience.

- Pull out 3 key statistics I can use in a presentation.

- What are the risks mentioned in this report?

- Give me a 100-word abstract of this article.

- Turn this research paper into a simple explanation for a high school student.

- List the top 3 recommendations from this document.

What once took hours can now take minutes, and you still retain control. AI helps you dive deeper into any section you need, but the AI helps cut through the noise.

CREATING A WEEKLY SCHEDULE OR ROUTINE

Planning a week can sometimes be overwhelming with so many meetings, deadlines, family commitments, errands, and workouts. Most people either overstuff their schedule or underplan and end up reactive. AI can help you find the balance by turning a list of commitments into a structured plan. All you need to do is feed the AI with your commitments and preferences, and it creates a schedule that includes tasks, breaks, meals, and downtime.

Suppose you have three meetings, two deadlines, and a soccer game to attend for your child. You type into the AI tool: *Here are my commitments this week [list them]. Create a realistic weekly schedule that includes exercise 3 times, dinner with my family each night, and at least 1 hour of reading for fun.*

The AI will produce a draft schedule, broken down day by day. You can then adjust times and priorities, but the hardest part, which is creating the structure, is already done. Other prompt ideas you can use for scheduling include the following:

- Create a morning routine that helps me be more focused at work by 9 a.m.

- Plan a balanced weekly schedule for someone working full-time, with 3 gym sessions and 2 hours for hobbies.

- I want to dedicate 5 hours this week to learning Spanish. Suggest where to place those sessions in my calendar.

The power is not just about saving time; it goes beyond that. It is about reclaiming mental clarity, allowing you to see your week laid out in a way that it feels achievable.

CREATING A TASK BREAKDOWN

Big projects can feel overwhelming, but writing "Finish Project X" on a task list does not help. It is too vague, and your brain will ultimately resist starting. The key to productivity is breaking large goals into smaller, specific steps. AI excels at this, and all you need to do is tell it what you are working on, and it will generate a task breakdown, often with suggested timelines.

Let's say you are tasked with launching a new marketing campaign. Many people would feel intimidated, but with AI, you type in: *I need to launch a social media campaign for our new product coming out in 6 weeks. Break this into weekly tasks, with milestones for planning, content creation, scheduling, and tracking results.* The AI might produce something like this:

- **Week 1:** Define goals, identify the target audience, and research competitors.

- **Week 2:** Draft campaign messaging and content ideas.

- **Week 3:** Create visuals and copy for approval.

- **Week 4:** Schedule posts across platforms.

- **Week 5:** Launch campaign and monitor engagement.

- **Week 6:** Collect results, analyze, and present findings.

This helps you achieve clarity and reduce procrastination, as the first step is now small and actionable, rather than vague and intimidating. Here are a few other prompts you can try out:

- *Break down the steps to plan a family vacation to Italy in 10 days, including booking flights, hotels, and sightseeing.*

- *Help me organize the process of writing a 20-page research paper with milestones over four weeks.*

- *List the steps for decluttering my garage in one weekend.*

These tasks you have seen might sound like small wins, but together they form the backbone of productivity. Most of the stress in our lives does not come from the big goals, but from the endless little decisions required to get there: *Do I read this now or later? Where do I fit this task into my week? What is the next step for this project?*

Each of those tiny questions chips away at our energy. AI helps by taking that work and organizing it, allowing you to feel lighter, calmer, and less stressed. To get the most out of this section, try this mini-experiment:

1. Pick one long document you have been avoiding. Paste it into an AI assistant and ask for a summary.

2. Write down all your commitments for next week. Ask AI to build a draft schedule.

3. Choose one big project or goal. Ask AI to break it into smaller steps with a timeline.

Notice how you feel afterward. The information is clearer, the week looks less overwhelming, and the big project suddenly feels manageable. That feeling you have, from chaos to clarity, is what makes AI such a powerful assistant.

QUICK PROMPT TEMPLATES FOR DAILY USE

The secret to efficiently using AI is knowing what to ask. Prompts don't need to be written perfectly, but the clearer and more specific you are, the better results you will unlock. This section will be your "cheat sheet," with ready-to-use prompt templates you can copy, paste, and adapt for your own life. To make them as flexible as possible, they have been written with placeholders with square brackets that look [like this]. Swap in your details, and you will have AI working for you in minutes.

SUMMARIZING ANYTHING QUICKLY

- *Summarize this article in 5 bullet points that highlight the main findings.*

- *Turn this 10-page report into a 200-word executive summary for busy leaders.*

- *Create a plain-English explanation of this research paper for someone with no background in [field].*

- *Give me a 30-second spoken summary of this [article/email] as if you were briefing me before a meeting.*

- *Highlight 3 action steps from this long document that I should focus on right away.*

WRITING AND EDITING EMAILS

- *Write a polite email to [name] requesting a meeting next week. Offer 2 possible times.*

- *Draft a professional but warm reply to [client name], updating them that the project will be done by [date].*

- *Rewrite this email to sound more concise and confident [paste draft].*

- *Create a friendly reminder email for [team/group] about [event/deadline]. Keep it short but clear.*

- *Write a thank-you email to [name] for their help with [project]. Include one specific detail about what I appreciated.*

CREATING WEEKLY SCHEDULES

- *Here are my commitments for the week [list]. Build me a daily schedule with time for breaks and meals.*

- *Design a balanced work-from-home routine that keeps me productive but avoids burnout.*

- *Plan a weekly routine for a parent with [number] kids, balancing work and family time.*

- *Suggest the best way to fit in 5 hours of exercise and 3 hours of study this week.*

- *Reorganize my commitments, so I have at least 1 free evening for relaxation.*

TASK BREAKDOWN FOR PROJECTS

- *Break down the steps for finishing [project] by [deadline] with weekly milestones.*

- *Turn "organize my house" into a step-by-step process broken down by room. Here is a home description [describe].*

- *Give me a daily task plan to write a 20-page research paper in 3 weeks.*

- *List the steps for hosting a dinner party for 8 people, starting 1 week in advance.*

- *Help me plan the launch of [product/event] with tasks divided into preparation, launch, and follow-up.*

BRAINSTORMING IDEAS

- *Generate 10 creative ideas for blog posts about [topic].*

- *Suggest 5 unique fundraising ideas for a small nonprofit.*

- *Give me 10 product name ideas for a [type of product].*

- *List creative ways to celebrate [holiday] with coworkers on a budget.*

- *Brainstorm 5 icebreaker activities for a group of [students/adults/professionals].*

DECISION-MAKING SUPPORT

- *List the pros and cons of [decision].*

- *Compare [option A] vs [option B] for [context]. Highlight cost, effort, and long-term benefits.*

- *If my priority is [goal], which option would make the most sense and why?*

- *Suggest 3 alternative solutions I may not have considered for [problem].*
- *What are the top 5 risks of choosing [option A], and how could I reduce them?*

MEAL PLANNING AND SHOPPING

- *Plan 5 dinners I can make in under 30 minutes. Include a shopping list.*
- *Create a vegetarian meal plan for 1 week with protein-rich recipes.*
- *Suggest easy lunchbox ideas for kids that are healthy and appealing.*
- *Make a 3-day meal plan using only these ingredients [list items].*
- *Give me snack ideas for someone trying to eat less sugar.*

LEARNING AND STUDYING

- *Explain [concept] to me in simple terms as if I were 12 years old.*
- *Create 10 flashcards for [topic], formatted as Q&A.*
- *Summarize this textbook chapter into 5 key points.*
- *Give me 3 real-world examples of [concept] in action.*
- *Turn this complex article into a study guide with key terms and definitions.*

PERSONAL WRITING HELP

- *Rewrite this paragraph for clarity and conciseness: [paste text].*
- *Polish this email to sound professional but approachable: [past text].*
- *Proofread this essay for grammar errors and suggest improvements.*

- *Make this LinkedIn post more engaging without changing the meaning: [paste draft].*

- *Shorten this message to under 100 words without losing key details.*

TRAVEL PLANNING

- *Plan a 3-day itinerary in [city], focusing on [interests: food, history, art].*

- *Create a budget-friendly travel plan for [destination] with estimated costs.*

- *Suggest a packing list for a 1-week trip to [destination] in [season].*

- *Find five family-friendly activities in [city].*

- *Plan a romantic weekend getaway in [location], including food, activities, and relaxation.*

FITNESS AND WELLNESS

- *Create a 30-minute no-equipment workout I can do at home.*

- *Design a weekly yoga routine for beginners.*

- *Suggest mindfulness practices I can do in under 10 minutes each morning.*

- *Write a daily habit tracker template for hydration, exercise, and sleep.*

- *Build a progressive strength training plan for 8 weeks.*

EVENT PLANNING

- *Plan a birthday party for a [age]-year-old who loves [theme].*

- *Write a checklist for hosting Thanksgiving dinner for [number] guests.*

- *Suggest icebreaker activities for a team-building event.*

- *Plan a graduation celebration with food, decorations, and music ideas.*

- *Draft an agenda for a 90-minute strategy meeting with time for discussion and decisions.*

PERSONAL REFLECTION AND JOURNALING

- *Give me 5 journal prompts to reflect on my week at work.*

- *Help me write a personal affirmation about confidence in [situation].*

- *Suggest end-of-the-day reflection questions to track mood and productivity.*

- *Create 5 gratitude prompts I can use each morning.*

- *Write a journal exercise to process stress about [issue].*

EXPANDED CHEAT SHEET: HOW TO BUILD GREAT PROMPTS

Prompts do not have to be fancy, but they do benefit from clarity. Here is a mini framework you can use to write your own:

1. **Task:** What do I want? (summarize, draft, brainstorm, or plan)

2. **Context:** What is it about? (an article, a project, or a meal)

3. **Details:** Who is it for? (boss, family, or myself)

4. **Format:** How do I want it? (bullet points, 200 words, or a schedule)

5. **Tone (optional):** What voice should be used? (professional, friendly, funny, or simple)

Think of it as ordering at a restaurant. If you just say "food," you will get something random. If you say, "I'd like a grilled chicken salad with no onions," you are more likely to get what you want. The more specific you are, the better the AI performs. Even if it does not get it right the first time, treat it like a conversation to guide it.

ACTIVITY: SAVE TWO HOURS THIS WEEK WITH AI

Here is where things get practical: You have seen what AI can do, but the best way to build confidence is by experiencing the payoff in your own life. This activity challenges you to reclaim at least two hours per week with AI. Treat it as an experiment where you will pick a few everyday tasks, hand them over to an assistant, and compare the results against doing it yourself. The outcome will be less stress, more time, and a stronger sense of just how powerful AI can be for ordinary life.

STEP 1: PICK TWO OR THREE EVERYDAY TASKS

Start by looking at your week ahead. Where do you feel stuck or slowed down? Common time-drainers include

- sorting through long notes or documents,

- drafting announcements or reminders,

- coordinating schedules with others,

- brainstorming options (meals, gifts, ideas), or

- turning vague goals into step-by-step tasks.

Pick two or three that feel most relevant. They do not need to be big or glamorous, as you will soon see that small wins add up fast.

STEP 2: CHOOSE A TOOL

Open one of the free AI assistants you learned about earlier and save it to your bookmarks bar if you haven't already. This makes it feel less like "researching a new app" and more like reaching for a calculator: It is just there, ready when you need it.

STEP 3: USE THESE FRESH PROMPT TEMPLATES

Here are some prompts you can try right away:

- Meeting prep

  - *I have a meeting with [person/team] about [topic]. Suggest 5 questions I should be ready to ask.*
  - *Turn these messy notes into a clean agenda for tomorrow's meeting.*

- Daily efficiency

  - *Plan a 10-minute morning routine that energizes me before work.*
  - *List 3 ways I can cut 20 minutes from my commute or morning prep time.*

- Household tasks

  - *Create a weekend cleaning checklist for a 2-bedroom apartment.*
  - *Organize these grocery items into a logical shopping order: [place list].*

- Personal growth

  - *Suggest a 30-day habit tracker template for drinking more water and exercising.*

  - *Give me 3 motivational quotes about focus I can put on sticky notes at my desk.*

- Quick writing help

  - *Draft a friendly text inviting a neighbor to a casual dinner.*
  - *Write a 2-sentence update I can post to my team's group chat about [topic].*

These prompts may look simple, but each of them is designed to trim off the minutes you normally spend thinking, drafting, and organizing.

STEP 4: COMPARE BEFORE AND AFTER

Estimate how long these tasks would normally take. Would you have spent 15 minutes writing that text, second-guessing every word? Ten minutes creating a shopping list? An hour building agenda? With AI, those tasks take seconds to generate and just a few minutes to edit. Add up the savings, and even if you only shave 10 minutes off 5 different activities, you have reclaimed nearly an hour. By combining scheduling, planning, and writing, the 2-hour goal is easy.

STEP 5: REFLECT ON THE WIN

At the end of the week, jot down your reflections:

- Which tasks felt easiest to delegate to AI?

- Which results surprised me?

- Where did AI save me the most time or stress?

The beauty of this activity is not just the two hours saved, but also the way your perspective will shift. Once you see AI clear away the repetitive clutter, you begin to trust it. However, you will also notice that with this discovery and trust comes a new curiosity: *What else can I offload?* That curiosity leads to even bigger wins.

JOURNAL PROMPT

You have spent this chapter exploring how AI can act as your personal assistant, and by now, you should already be able to see how the small wins add up. Maybe you have already saved time, perhaps you already feel less stressed, or you may be surprised by how natural it feels to talk to a digital assistant. Now it is time to pause and reflect. Pick up your notebook or open a blank page on your computer and take a few minutes to write on these two prompts:

What did AI do well for me this week?

Think of a specific moment where you used AI. It could be drafting a message, organizing your tasks, or helping you see a big project more clearly. Describe both what it did and how it made you feel. Did it save time? Relieve pressure? Help you feel more in control? The more details you include, the more real the win will feel.

Where did AI struggle?

No tool is perfect, and noticing the rough edges is part of learning. Maybe AI's draft email sounded too formal. Perhaps the summary missed something important. The schedule it suggested may feel unrealistic. You should see these as opportunities, not failures. By spotting the limitations, you are learning how to steer the tool more effectively.

Finally, write one more sentence:

Next week, I want to try using AI to...

This is your personal challenge. It might be another productivity task, like organizing your files or prepping meeting notes. It can even be something outside this chapter's scope. What matters is that you identify the next step where AI could make your life easier.

So far, you have seen AI act like a behind-the-scenes assistant. It has helped you shave hours off repetitive work and clear away the clutter that eats up your energy. But productivity is only one side of the story, as you are about to see.

The real transformation happens when AI helps you grow, and not only saves time. Imagine having someone who not only keeps your calendar but also tutors you on a new subject, brainstorms creative ideas with you, or helps you unlock skills you did not know you had.

You have already seen bits and pieces of this, but in the next chapter, we will go further. You will discover how AI can be your personal study partner, a brainstorming collaborator, and even a spark for creativity when you are stuck. If Chapter 4 was about clearing space, Chapter 5 is about filling it with knowledge, growth, and imagination.

CHAPTER 5:

SUPERCHARGING LEARNING AND CREATIVITY

THERE IS A QUIET MOMENT MOST LEARNERS AND CRE-ators know well: staring at a blank page, an unopened textbook, or a half-finished idea that feels like it is going nowhere. Although it might seem like laziness, most of the time it is the weight of information overload, the fear of not being good enough, or even the exhaustion of managing too many responsibilities. In that moment, the joy of learning or creating feels far away, replaced by frustration and self-doubt.

That is where AI can become a powerful ally. However, it is not because it swoops in with a ready-made answer or finished product. It is because it gives you momentum. It helps you see the first step, nudges you past the blank page, and reminds you that learning and creativity don't have to be struggles fought alone.

One student of mine, during a conversation about AI, revealed that although they did not use ChatGPT for essays, they used it to break down the question so they could actually understand what they needed to do. They said it felt like having a patient tutor who never got tired of their questions. Another chimed in and stated that whenever they got stuck with a song they were working on, they would ask the AI for chord progression suggestions to continue, as if they were bandmates sharing ideas.

These voices highlight an important truth: AI does not replace learning or creativity; it enhances them. The ideas and sparks are still yours, and the AI will simply lower the barrier that keeps you from getting there. You are still doing the work, but with less struggle and more confidence. The difference can mean the gap between giving up on an idea and bringing it to life.

For learners, the relief is immediate, where instead of spending hours confused by a dense article, they can ask AI: *Summarize this in plain English.* Instead of staring at a math problem until your head hurts, you can say: *Explain this step-by-step as if I were in high school.* The answer is not meant to bypass your effort, but to guide it. With clearer explanations, you have more energy left for practice, application, and deeper understanding.

For creators, the benefit is just as clear. The black canvas is intimidating, whether it is a document, a sketchpad, or a music track. AI can generate the first draft you need by brainstorming 10 titles for your blog post, offering rough layout ideas for a flyer, or suggesting a hook for a song lyric. You do not have to use what it gives you word-for-word; you can mix, tweak, and adapt until it feels right. The key here is that you are no longer stuck at zero.

When you start using AI with intention and purpose, you will see something unexpected happen: working with AI often makes people more creative, not less. By removing the starting paralysis, it frees up mental space to play. When you are no longer looking to get it right, you can explore, experiment, and imagine more boldly. This is what makes AI a powerful partner: It restores the sense of joy. The joy of curiosity without the fear of confusion and the pressure of perfection. It is the joy of moving forward instead of staying stuck.

You do not have to be a professional artist, musician, or academic to benefit. You just have to be someone who sometimes feels overwhelmed, stuck, or uninspired, which, in other words, means all of us. AI is here to help you discover the joy of work, and it is exactly this that you will learn in this chapter.

LEARN ANYTHING FASTER WITH AI

We live in an age where information is available everywhere and anywhere, but that does not make learning any easier. Sometimes, the problem is not access to knowledge, but figuring out how to approach it without getting overwhelmed by details or becoming discouraged. This is where AI makes a difference as a study partner, as it helps structure your learning, tests your knowledge, and explains complex ideas in a way that makes sense.

Here are five real-world case studies showing how everyday learners can supercharge their education with AI, plus practical prompts you can adapt for your own purposes.

CASE STUDY 1: CUSTOM STUDY PLANS

Olivia, a nursing student, was overwhelmed by the volume of material she needed to review before her exams. She knew she could not memorize everything at once, but she did not know how to pace herself. She asked an AI assistant:

> *Create a 4-week study plan to prepare for my nursing exams. I need to cover anatomy, physiology, and pharmacology. Schedule shorter reviews on weekdays and longer sessions on weekends.*

The AI broke down her goals into manageable daily chunks, complete with reminders to review old material. Olivia followed the plan, and instead of cramming, she felt steady progress and reduced stress.

Extra prompts to try:

- *Design a 7-day plan to learn the basics of [topic]. Include 30 minutes of study every day.*

- *I have 2 hours tonight. Suggest the best way to review [topic] effectively.*

- *Turn this syllabus into a weekly schedule I can realistically follow.*

CASE STUDY 2: FLASHCARDS AND QUIZZES

David, a high school sophomore, was struggling to memorize key terms for his history class. Reading the textbook over and over was not sticking. He copied key terms into his AI tool and asked the following:

- *Create 20 flashcards with the term on 1 side and the definition on the other.*

- *Turn these notes into a multiple-choice quiz with answers.*

AI instantly generated practice materials that David used daily and found himself recalling information much faster.

Extra prompts to try:

- *Create 10 fill-in-the-blank questions about [topic].*

- *Make a short-answer quiz for [chapter or lesson].*

- *Turn this article into 15 flashcards with simple definitions.*

CASE STUDY 3: EXPLAIN THIS TO ME LIKE...

Priya, a marketing professional, wanted to learn basic coding but felt lost in jargon. Tutorials felt too technical, and she almost gave up. Instead of asking for a technical definition, she tried:

- *Explain what a variable is in programming as if I were a 10-year-old.*

- *Now explain it as if I were a college student learning for the first time.*

- *Finally, explain it as if I am preparing for a job interview in programming.*

The layered explanations gave Priya the simple understanding she needed to get started and the professional context she needed to grow.

Extra prompts to try:

- *Explain [concept] in three different ways: for a beginner, an intermediate learner, and an expert.*

- *Use an analogy from sports/music/cooking to explain [concept].*

- *Break this problem into steps, and explain each step in plain English.*

CASE STUDY 4: LANGUAGE LEARNING PARTNER

Marco was learning Spanish, but he had no one to regularly practice with. Reading was fine, but speaking felt awkward. He asked an AI tool:

- *Pretend you are a Spanish tutor. Ask me five beginner-level questions and correct my answers.*

- *Give me 10 practice sentences in Spanish using the past tense, then translate them.*

- *Have a simple conversation with me in Spanish about ordering food in a restaurant.*

Suddenly, Marco had a conversation partner available 24/7, without judgment. His confidence grew with each practice.

Extra prompts to try:

- *Translate these sentences into [language] and explain any grammar rules used.*

- *Quiz me on 15 new vocabulary words for [topic].*

- *Correct my mistakes in this paragraph I wrote in [language].*

CASE STUDY 5: PROFESSIONAL SKILL GROWTH

Jasmine, a project manager, wanted to sharpen her leadership skills. She did not have time for a course, but she wanted a targeted price. She called AI to the rescue and asked:

- List 10 common challenges project managers face and suggest strategies to handle them.

- Create a role-play scenario where I need to give constructive feedback to a team member.

- Summarize the 3 most important leadership styles and when to use them.

These AI-guided exercises became a part of Jasmine's weekly routine. Within months, she noticed she was handling workplace challenges with more confidence.

Extra prompts to try:

- *Give me a case study about [professional skill] and ask me how I would handle it. Then give me feedback.*

- *Summarize the top 5 books on [topic] and extract their key ideas.*

- *Generate 10 interview-style questions about [skill] so I can test myself.*

These five case studies highlight the core strength of AI as a learning partner: adaptability. Whether you are a student, a professional, or simply curious, you can shape the tool to fit your goals and your pace. It does not always replace the effort you need to put in, but it amplifies it, allowing you to clear confusion and give you practice in a way that feels engaging. The secret is in the prompts. The more you ask AI to adapt, the more it feels like a personal tutor designed just for you.

CREATIVE BOOSTS: WRITING, ART, AND MUSIC

If learning with AI feels like having a personal tutor, then using AI for creativity feels like inviting a brainstorming partner who never runs out of ideas. Whether you are writing a blog post, designing a flyer, or experimenting with music, AI can take the edge off the blank page (or canvas, or soundboard) and help you unlock your imagination.

The key is to remember that AI is not here to replace your creative spark, but to ignite it. The final choices, the vision, and the artistry are still yours. Let's explore three of the most common creative areas where AI can give you a boost: writing, visuals, and music.

WRITING WITH AI

Writers often say the hardest part is getting started. A blank page can feel intimidating, but AI can jumpstart the process with outlines, drafts, and fresh ideas. From professional writing to personal journaling, its applications are near endless. Here are three ideas to help you get started:

1. **Blog posts and articles**

   o *Give me 10 headline ideas for a blog post about [topic].*

   o *Draft a 500-word blog post on [topic] with a friendly, approachable tone.*

o *Create an outline for a how-to guide about [skill] with clear steps.*

o *Summarize the 3 most recent trends in [industry] and explain their impact.*

o *Generate 5 attention-grabbing introductions for a blog post about [topic].*

2. Storytelling

o *Create 3 character profiles for a mystery story set in [location].*

o *Suggest 5 possible endings for a story about [theme].*

o *Write the 1st paragraph of a story about someone who discovers [situation].*

o *Give me dialogue ideas for 2 characters arguing about [conflict].*

o *Help me write a plot outline for a children's bedtime story involving [animal or object].*

3. Social media posts

o *Write 10 Instagram captions for photos of [type of content]. Use a playful tone.*

o *Suggest a 1-week content calendar for LinkedIn posts about [topic].*

o *Draft -5 X posts [event/product]. Mix humor and insights.*

o *Write a short, friendly Facebook post announcing [news].*

o *Create 3 TikTok script ideas under 30 seconds about [topic].*

As you will see, there will be great starting points for your projects, allowing you to tweak the content and still maintain your voice. All AI will do is help you test different tones, formats, and ideas until one sticks, and you can develop on it.

VISUAL CREATIVITY WITH AI

Not everyone can draw or design, but everyone has ideas. AI art tools like Canva's Magic Studio, Dall·E, and NightCafe allow you to turn words into images in seconds. Here are a few useful prompts to adapt and use for your own purposes:

1. **Quick design projects**

   o *Create a flyer design for a community yoga class. Include calming colors and simple text.*

   o *Generate a presentation cover slide for a business meeting about [topic]. Use a professional style.*

   o *Design a minimalist logo for a coffee shop called [name].*

   o *Give me 3 poster ideas for a fundraiser event for [cause].*

   o *Suggest layouts for an infographic about [topic].*

2. **AI-generated art**

   o *Generate a surreal painting of a city floating in clouds.*

   o *Create an illustration of a dragon reading a book in a library.*

   o *Design a realistic image of a futuristic classroom using AI assistants.*

   o *Draw a cartoon-style character based on this description [insert details].*

   o *Turn the phrase "hope into chaos" into a symbolic digital artwork.*

3. **Everyday visuals**

   o *Make a birthday invitation for a 10-year-old who loves dinosaurs.*

   o *Design a thank-you card with a cheerful, hand-drawn style.*

o *Create a simple social media graphic announcing a sale on [product].*

o *Generate a background image for a Zoom meeting that looks like a cozy library.*

o *Produce a minimalist calendar layout for the month/year.*

Here, you will see that accessibility is one of the greatest advantages of AI tools. You will be able to edit and make changes to the illustration so that it is exactly like you imagined. You do not need to be a designer; all you need is a vision so that AI can handle the execution. You can just focus on refining until it feels right.

MUSIC AND SOUND WITH AI

Music might feel like the most untouchable form of creativity, but with AI tools like Suno AI, Soundraw, and Amper Music, the process is remarkably approachable. You do not have to be a professional musician to experiment with soundscapes, melodies, or lyrics. Use these prompts to help you bring your musical talent to the next level:

1. **Generating music ideas**

   o *Compose a 30-second upbeat tune for a podcast intro.*

   o *Generate a calm, ambient soundtrack for meditation.*

   o *Create a background track in the style of 1980s synthwave.*

   o *Suggest 5 chord progressions for a pop song in C major.*

   o *Write lyrics for a chorus about [theme].*

2. **Adding chords or structure to lyrics**

   o *Suggest chord progressions for these lyrics: [paste lyrics].*

   o *Turn this poem into a song by suggesting verses and a chorus.*

   o *Provide drum patterns for a song in 4/4 time.*

   o *Generate harmonies for this melody line: [describe or paste notes].*

    ○   *Suggest 3 different arrangements (acoustic, electronic, and orchestral) for this song idea.*

3. Sound design for projects

    ○   *Generate sound effects for rain, thunder, and footsteps on gravel.*

    ○   *Create a background track for a documentary about nature.*

    ○   *Design a suspenseful audio loop for a video game scene.*

    ○   *Provide a relaxing piano instrumental for studying.*

    ○   *Make a short jingle for a local bakery advertisement.*

Even if you do not consider yourself "musical" or "creative," AI makes experimenting fun. It takes the pressure off and turns the creative process into play. These may seem like different worlds, but they share a common challenge: starting. AI helps break that inertia and gives you enough momentum to keep moving.

AI can work as your ultimate brainstorming buddy by throwing ideas at you. Some will be good, some will be bad, and some will be surprising. You get to decide which ones to keep and which ones to throw out. That freedom helps your creativity flourish, and as a reminder that you do not have to create alone.

ACTIVITY: CREATE A SEVEN-DAY LEARNING OR CREATIVE CHALLENGE

One of the best ways to build confidence with AI is to use it daily, even for small tasks. By making it a routine, you will save time and discover how versatile it can be for learning and creativity. This is what this challenge is all about.

Over the next seven days, I propose that you follow this activity to increase your skills with AI. You should pick a skill or a creative medium you would like to explore, something you have been curious about but have not had the time or confidence to dive into. It could be writing short stories, sketching, learning a language, exploring photography, or even songwriting. Then, with AI as your creative partner, you will tackle a small exercise each day. By the end of the week, you will have new skills *and* proof that AI can amplify your creativity and make learning more enjoyable.

STEP 1: CHOOSE YOUR FOCUS

Start by picking one of these categories, or invent your own:

- Writing: short stories, essays, poems, or blog posts

- Visual design: posters, social media graphics, and character sketches

- Music: lyrics, melodies, soundscapes, or instrument practice

- Language learning: vocabulary, dialogues, and cultural knowledge

- Knowledge building: science, history, business, or any subject you want to master

Do not overthink it and choose something that excites you. The goal is not perfection, but exploration.

STEP 2: SET A SMALL GOAL

Frame your challenge with a clear outcome, such as

- *I want to write 1 short story draft by the end of the week.*

- *I want to create 3 pieces of digital art.*

- *I want to learn 30 new Spanish vocabulary words.*

- *I want to understand the basics of climate change science.*

This will give you direction and allow you to practice for about 20–30 minutes per day.

THE SEVEN-DAY PLAN

Here is a template you can follow by swapping in your focus and adapting the prompts to match your goal.

Day 1—Brainstorm with AI: The objective here is to generate a pool of ideas from which you can draw inspiration and the initial ideas for your project.

- Writing: *Suggest 10 short story ideas about friendships in unusual places.*

- Visuals: *Generate 10 creative poster ideas for a festival in a futuristic city.*

- Music: *List 10 themes for a song about resilience.*

- Language: *Give me 20 basic phrases in French for ordering food.*

- Knowledge: *List the 10 most important concepts in [subject].*

- At the end of the session, pick 2–3 ideas that excite you the most.

Day 2—Learn the basics: The objective is to gather foundational knowledge and get a start on your project. AI will act like your tutor or coach and guide you through the creation process.

- Writing: *Create an outline for a short story based on idea #3. Include beginning, middle, and end.*

- Visuals: *Suggest a step-by-step guide to sketch a character from idea #2.*

- Music: *Explain 3 common chord progressions for pop songs.*

- Language: *Create a vocabulary list with example sentences for 10 useful phrases.*

- Knowledge: *Explain [concept] as if I were a beginner, in 200 words.*

Day 3—First draft or attempt: Here, you will make your first attempt at creating something simple from the ideas you have had so far.

- Writing: *Write the first 300 words of the story outline from yesterday.*

- Visuals: *Generate a poster draft in Canva for the festival.*

- Music: *Compose 4 lines of lyrics based on the resilience theme.*

- Language: *Create a dialogue between a traveler and a waiter using yesterday's vocabulary.*

- Knowledge: *Summarize the key arguments about [topic] in 5 bullet points.*

Day 4—Feedback and iteration: Use AI to refine your first attempt and adjust it to your vision.

- Writing: *Rewrite this draft to make the dialogue more natural.*

- Visuals: *Suggest 3 variations of this poster design with different color palettes.*

- Music: *Suggest 2 chord progressions to do with these lyrics.*

- Language: *Correct my dialogue for grammar and suggest improvements.*

- Knowledge: *Explain where my understanding of [topic] might be incomplete.*

Day 5—Expand and experiment: Use this opportunity to push your creativity further and experiment with stretches on your imagination.

- Writing: *Add a plot twist in the middle of this story.*

- Visuals: *Turn this sketch into 3 different artistic styles: comic, watercolor, and surreal.*

- Music: *Transform my song lyrics into a country style, then into a rock style.*

- Language: *Write a short story using 15 of my new vocabulary words.*

- Knowledge: *Explain [concept] in a cooking metaphor.*

Day 6—Polish and prepare: Move toward a finished product, where AI will help you shine your rough idea into something you can share.

- Writing: *Polish this story draft so it is clear, concise, and engaging.*

- Visuals: *Create a final, high-resolution version of my poster with clean typography.*

- Music: *Arrange this song idea into a 1-minute demo with intro, verse, and chorus.*

- Language: *Quiz me with 10 questions using the vocabulary I have learned this week.*

- Knowledge: *Turn my notes into a 1-page cheat sheet on [topic].*

Day 7: Showcase and reflect: Share your creation and reflect on the process.

- Writing: Post your short story draft in a writing group or read it aloud to a friend.

- Visuals: Share your design on social media or print it out.

- Music: Play your demo for a friend or record yourself performing it.

- Language: Hold a short conversation with a friend or practice aloud.

- Knowledge: Teach what you have learned to someone else using your cheat sheet.

Finally, ask yourself:

- What did AI help me with the most?

- Where did I add the most personal creativity?

- What will I try next week?

In a week, you will have proof that AI can save time and help you learn, grow, and create in ways you might not have thought possible.

EXTRA PROMPTS TO TRY OUT

One of the best ways to explore AI is through play. Instead of waiting until you "need" it, give yourself the freedom to experiment. Below you will find a new set of prompts designed to flex your imagination, make learning more engaging, and spark creativity in unexpected directions. They are flexible, so you can adapt them to whatever topic or project excites you the most.

- **Speed learning**
  - *Explain the history of [event] in under 2 minutes, as if you are making a YouTube Shorts Script.*

- o *Summarize the 5 most important things to know about [topic] if I only had one day to study.*

- o *Turn [concept] into a short poem that helps me remember it.*

- o *List 3 surprising facts about [topic] that most beginners do not know.*

- o *Explain [theory] through a real-world scenario involving grocery shopping.*

- **Active recall**

  - o *Quiz me with 10 true/false questions about [topic]. Don't give me answers until I guess.*

  - o *Ask me 5 increasingly harder questions about [concept], like a mini-game.*

  - o *Pretend you are an interviewer asking me about [subject]. Challenge me with tough follow-ups.*

  - o *Give me a quick case study about [skill] and ask me how I would handle it. Wait for my response before providing feedback.*

  - o *Turn this passage into fill-in-the-blank questions with the answers hidden.*

- **Unusual story starters**

  - o *Begin a story with the line, "This morning, I woke up, and the sun was gone..."*

  - o *Write the opening to a story where the narrator is a houseplant observing humans.*

  - o *Start a mystery with the discovery of a locked journal in an attic.*

  - o *Write the first 200 words of a comedy about a professional ghost hunter who is terrified of the dark.*

  - o *Imagine a world where gravity works differently. Write a short scene.*

- **Nonfiction and persuasion**
    - *Draft a 2-paragraph opinion arguing why [everyday object] is underrated.*
    - *Explain the pros and cons of [topic] as if writing for a school newspaper.*
    - *Turn my bullet-point notes on [subject] into a clear, flowing article.*
    - *Write a motivational email encouraging someone to stick with [habit].*
    - *Give me 5 taglines for a campaign about [cause].*

- **Playful social writing**
    - *Invent a fake but funny product and write a short advertisement for it.*
    - *Write 5 witty 1-liners about [topic] I could use as icebreakers.*
    - *Create a horoscope-style prediction for someone born today.*
    - *Write a 4-line poem formatted as a tweet about [topic].*
    - *Pretend you are a historical figure reacting to today's technology. Write their post.*

- **Creative illustration**
    - *Draw an animal that does not exist but could plausibly live in a rainforest.*
    - *Create a travel poster for a city that only exists in a fantasy novel.*
    - *Design a children's book illustration showing [animal] learning to ride a bicycle.*
    - *Make a surreal landscape where the sky is made of books.*
    - *Generate a portrait of a time traveler lending in during three different historical periods.*

- **Functional visuals**
    - *Design a resume layout for a creative professional.*

- o *Create icons for a productivity app, each symbolizing a different task.*
- o *Generate a visual mood board for a cozy café aesthetic.*
- o *Design 3 different packaging ideas for eco-friendly soap.*
- o *Make a simple step-by-step infographic for how to brew the perfect cup of tea.*

- **Song experiments**
  - o *Write a lullaby about outer space.*
  - o *Create a rap verse explaining [scientific concept].*
  - o *Suggest lyrics for a duet between 2 unlikely characters: a robot and a tree.*
  - o *Generate a chorus for a rock anthem about perseverance.*
  - o *Turn a famous proverb into a catchy song lyric.*

- **Atmosphere and landscapes**
  - o *Design ambient background music for a rainy café.*
  - o *Create a suspenseful sound design for a short horror film scene.*
  - o *Generate cheerful carnival music with brass instruments.*
  - o *Compose a meditation soundtrack that uses ocean sounds as rhythm.*
  - o *Produce a theme song for a podcast about technology and human stories.*

HOW TO GET THE MOST OUT OF THESE PROMPTS

These prompts will act as springboards to help you start your creative process using AI. The point here is not to create a polished final product each time, but to explore, experiment, and see what happens. Here are a few tips to maximize the experience:

- Mix and match by trying to learn at least one prompt per category, then move on to a visual or music prompt next.

- If the first results feel flat, push the AI further. Iterate until you get the result you were looking for, and that's all you need are small tweaks.

- Replace placeholders with your hobbies, interests, or current goals. A prompt about her outer space can just as easily become one about gardening or basketball.

- After each mini-experiment, jot down how it felt: Did it save time? Did it spark an idea it would not have otherwise?

By using these tools, you will leave this chapter with a deeper appreciation of AI as a learning and creative partner and with a growing library of your own personal prompts.

JOURNAL PROMPT

Take a breath. You have just completed one of the most exciting parts of this book. By now, you have already discovered how AI can be more than a time-saver or a background assistant. You have seen how it can ignite your creative spark and work as a private and customized tutor available at any time. But the real breakthrough is not using the prompts you used, but how you felt. Did you notice the shift? That moment when confusion turned into clarity, or hesitation turned into playfulness? Before we move on, let's capture that growth.

Grab your notebook or open a blank page and reflect on these questions:
What creative or intellectual barrier did AI help overcome?

Think about one specific example. Maybe AI explained a concept in a way that finally clicked. Maybe it generated a dozen creative ideas when you thought you had none left. Maybe it helped you finish something you had been procrastinating on for weeks. Capture the moment in detail: What were you struggling with, what prompt did you use, and how did it shift your perspective?

How did I feel using AI to learn or make something?

Was it relief? Excitement? Surprise? Maybe even a little discomfort? Be honest here. Reflecting on your emotional response helps you notice whether AI is lifting you up, overwhelming you, or sparking curiosity.

Where do I want to go next?

Write one sentence starting with: *Next, I want to use AI to...* Fill in the blank with whatever feels most natural. It may be improving a skill, exploring a new art form, or tackling a subject you have always avoided. This sentence becomes your personal north star for the next part of your AI learning experience.

Once you are done, it is time to move on. In the next chapter, you will explore how to apply AI more deeply to the way you learn and create. The chapter will be about using AI to help you generate income. As you will see, there are several possibilities to make money with side hustles using AI, all just a few keystrokes away. Let's take a look!

CHAPTER 6:

AI SIDE HUSTLES AND PASSIVE INCOME

FOR GENERATIONS, STARTING A BUSINESS WAS A RISKY, RE-source-intensive venture that required substantial funding for equipment, considerable time to build everything from scratch, and a team of specialized experts. These included designers for branding, copywriters for messaging, and marketers for outreach. The barrier to entry was high, which is why so many people with great ideas never took the leap.

Today, this scenario has shifted, where you can have a business partner who works for free, does not sleep, and can help you design logos, draft product descriptions, write blog posts, brainstorm business models, and even create marketing campaigns. This partner is AI. At the same time, this does not mean that AI does everything perfectly; let's be honest, it does not. However, when you pair it with your creativity, judgment, and passion, it can accelerate the process of turning an idea into a side hustle faster than most people realize.

Lisa was a teacher who loved organizing her classroom with creative worksheets. She had always thought about selling them online, but did not know where to start. Using AI tools, she generated polished, age-appropriate designs for math and reading worksheets. These were then packaged into PDFs and uploaded to an online marketplace like Teachers Pay Teachers. Within weeks, she had a new income stream, something she had dreamed about but never thought was possible without hiring a designer.

We then have James, a coffee enthusiast who decided to create a small online guide about brewing methods. Instead of spending months writing from scratch, he used AI to create a draft outline, expand sections, and polish the language. He then designed a simple ebook cover with a free AI tool and published it on Amazon Kindle Direct Publishing (KDP). His guide started earning sales from people just as obsessed with coffee as he was.

Finally, there is Sandy, a stay-at-home mom who turned her love of organizing into a microbusiness. With AI's help, she created editable planner templates with daily checklists, meal prep guides, and budgeting sheets, and listed them on Etsy. What started as a small experiment began bringing in steady side income, giving her more financial freedom without requiring a full-time job.

These individuals are not outliers who got lucky. There are just examples of people around the world who are tapping into AI to bring their skills, passions, and curiosities into the marketplace. The best part is that you do not need a big budget to try this out for yourself.

Many of the AI tools you will use are free or very low-cost. Where once you might have had to spend hundreds of dollars hiring someone to design a logo, you can now create one yourself in minutes. Instead of paying for market research reports, you can ask AI to scan the trends and suggest niches worth exploring. Instead of hiring a copywriter to draft your product descriptions or email campaigns, you can immediately generate polished drafts.

However, it is essential to remember that AI does not replace your judgment. It won't actually know your audience as deeply as you do, and it can't always generate accurately. That is why your role as the strategist and fact-checker, as well as the person who injects authenticity, is essential.

It is crucial to remember that while it is tempting to see AI as a magic machine, it is important to stay grounded. AI tools generate results based on patterns, not deep understanding. This means

- **Always fact-check.** If you are writing an informational guide, double-check any statistics or claims. AI can sometimes "hallucinate" facts.

- **Refine outputs.** The first draft might be generic. Your role is to edit, refine, and personalize it so it reflects your voice and values.

- **Avoid overreliance.** Remember that AI is a tool, not a business in itself. Sustainable side hustles still require customer empathy, testing, and iteration.

By approaching AI with both excitement and discernment, you will find that it becomes the business partner you needed to start your business. This is the promise of the side hustle era, powered by AI: You can start small now and start with almost nothing but your imagination. The process is simple: All you need is to find and validate the right idea.

FINDING AND VALIDATING IDEAS WITH AI

Every successful side hustle begins with a spark, an idea that connects what you can offer with what others want or need. For many people, this might feel elusive, especially if the "Where do I start with" is the part you have the most trouble with. The good news is that you do not need inspiration to strike while driving or in the shower. With AI, you can generate, explore, and validate potential ideas in a fraction of the time it once took.

At the same time, the internet is filled with lists of "side hustle ideas," but not all of them will fit your skills, interests, or lifestyle. Some sound promising, but may already be saturated. Others may be too niche to generate meaningful income. Instead of guessing, you can use AI as a research partner that scans patterns, identifies opportunities, and helps you test whether an idea has potential before you put your energy into it.

Imagine you are interested in fitness. Instead of typing "fitness side hustles" into a search engine and going through generic advice, you can ask AI, *What are three underserved niches in the online fitness space where digital products could succeed?* Within seconds, it might suggest things like workout plans for busy parents, beginner-friendly strength training for seniors, or home fitness routines for people with limited equipment. Already, you have moved from vague interest to specific, testable niches.

Once this is done, AI can help you look at what is trending. Tools like ChatGPT, Claude, or Gemini won't have real-time data like TikTok's trending hashtags or Amazon's bestseller lists (unless paired with browsing), but they can analyze general patterns, point to evergreen niches, and even suggest where rising interest is happening. When combined with your curiosity, these outputs can guide you toward markets where people are already spending time and money. For example:

- *List 10 trending niches in digital products for 2025 that align with growing lifestyle or health interests.*

- *What hobbies or topics are becoming popular among Gen Z audiences that could translate into digital content or products?*

However, finding an idea is one thing, and ensuring people will actually pay for it is another. AI cannot guarantee sales, but it can help you think through monetization models. You can ask the following:

- *What are 5 ways to monetize a blog about sustainable living?*

- *How do creators in the digital art space typically generate income from their work?*

These prompts help you quickly see if your ideas have legs or whether you may need to tweak them before investing more time.

TEN PROMPTS TO SPARK AND TEST IDEAS

Here is a starter pack of prompts you can try right away. Adapt them to your interests and background.

- *Generate 10 side hustle ideas for someone interested in [skill/topic].*

- *What underserved audiences exist in the [industry] market, and what products or services might they need?*

- *Summarize 3 ways people are currently monetizing [hobby/interest].*

- *Compare the earning potential of selling digital templates on Etsy versus launching a small ebook on Amazon.*

- *List 5 trending content topics on YouTube or TikTok that could be turned into digital products.*

- *What pain points do busy professionals have that could be solved with simple digital tools or resources?*

- *Give me 10 examples of tiny side hustles that require less than $50 to start.*

- *Pretend you are a customer in [audience type, ex. new parents]. What problems or needs would you pay to solve?*

- *Evaluate the pros and cons of starting a newsletter about [topic].*

- *Suggest ways to validate whether people would pay for a digital product about [idea].*

Each of these prompts moves you beyond vague brainstorming. They will help you generate ideas, highlight gaps, customer pain points, and potential income streams.

PUTTING IT INTO PRACTICE

Here is how you might use these prompts in action. Suppose you are a college student interested in photography. You try prompt #1: *Generate 10 side hustle ideas for someone interested in photography.* The AI suggests

- selling a preset filter for photo-editing apps.

- creating a beginner's guide to smartphone photography.

- offering simple event photography services.

- teaching photo composition through short e-courses.

- curating niche stock photo packs.

You like the idea of preset filters. Next, you use prompt #6: *What pain points do busy professionals have that could be solved with simple digital tools or resources?* The AI notes that many professionals want to share high-quality images on LinkedIn without spending hours editing. Suddenly, your idea has a clear audience.

Finally, you use the last prompt: *Suggest ways to validate whether people would pay for a digital product about photography presets.* The AI suggests setting up a small Etsy shop with a free sample pack, posting in photography communities to spark interest, and testing ads with minimal spend. Within minutes, you have moved from a general interest in photography to a niche side hustle idea, a defined audience, and a pathway to validation.

The hardest part of starting a side hustle is not always the work: It is the uncertainty of where to begin. AI helps cut through that uncertainty by giving you a systematic way to brainstorm, refine, and test ideas. You do not have to reinvent the wheel or wait for inspiration. You have to start asking the right questions.

CREATING DIGITAL PRODUCTS AND CONTENT WITH AI

Once you have validated an idea, the next step is bringing it to life. For many side hustles, that means creating digital products or content that can be packaged, marketed, and sold. This is where AI shines. With the right tools and prompts, you can accelerate what once took weeks or months into days or even hours.

Digital products are appealing because they are low-cost to create, easy to distribute, and infinitely scalable. You build them once, and they can be sold over and over without shipping or inventory. Content, on the other hand, is the fuel that attracts and engages your audience, whether through social media posts, blogs, or emails. Together, they create the backbone of most AI-powered side hustles. Let's break down three of the most popular categories and show how AI can help you master each.

BOOKS AND GUIDES

Self-publishing is one of the most accessible ways to monetize your knowledge or passion. Whether it is a short how-to guide, an ebook, or a workbook, digital publishing allows anyone to share their expertise and earn income without going through traditional publishers. In this, ebooks and guides work because they position you as an authority in your niche. They can also be sold on platforms like Amazon KDP, Gumroad, or Etsy. Finally, these can also be short, sometimes with 20 to 30 pages, as long as they solve a problem or provide value.

AI will help since, traditionally, writing a book will feel daunting because of the blank page problem. AI removes that barrier by helping you brainstorm book titles and outlines, generate first drafts for sections or chapters, suggest examples, analogies, and exercises, and polish text for clarity and flow.

Here is a sample process you can use:

Imagine you are passionate about meal prepping. Here is how you could use AI to create a guide:

1. **Outline generation:** You can use prompts such as, *Generate a detailed outline for a beginner-friendly guide on meal prepping for busy professionals. Include 8–10 chapters and list key points for each.*

2. **Drafting chapters:** Prompts such as, *Write a 1,000-word draft for Chapter 3, "Meal Prep on a Budget," including practical tips and at least 3 sample recipes.*

3. **Adding personal flair:** You should not just copy and paste, but edit, add your own voice, and include real-life tips.

4. **Designing the book:** Tools like Canva can turn your text into a polished PDF with attractive layouts and graphics.

5. **Publishing:** Upload to Amazon KDP, Gumroad, or Etsy and start selling.

Other prompts you can use for books and guides include

- *Brainstorm 10 book titles for a guide about [topic].*

- *Create a chapter-by-chapter outline for a how-to guide on [topic].*

- *Write a case study example for a section on [specific concept].*

- *Rewrite this draft in a more conversational tone suitable for beginners.*

- *Summarize the key lessons from Chapter 5 in a bulleted list for a workbook page.*

SOCIAL MEDIA CONTENT

The engine behind successful side hustles is usually social media. Whether you are selling a digital product, building a personal brand, or driving traffic to your website, consistent and engaging posts help you stay visible and relevant. The challenge, however, is that this task is time-consuming, which is where AI comes to the rescue.

In this case, AI will be able to generate post ideas and captions, rewrite text in different tones for different platforms, suggest hashtags and posting schedules, draft scripts for short-form videos like TikTok or Reels, and turn long content into bite-sized social posts. Let's say you have created a digital guide on "Mindful Morning Routines." You want to promote it on Instagram and LinkedIn. Here is how you could use AI:

1. **Generate ideas:** Try the following prompt: *Generate 15 Instagram post ideas to promote a guide about mindful morning routines. Include hooks, captions, and call-to-actions.*

2. **Create varied formats:** Prompts such as, *Rewrite this blog excerpt into a LinkedIn Update in a professional or approachable tone* or *Turn this section of my book into a 30-second TikTok script.*

3. **Visual design:** Tools like Canva Magic Studio or DALL·E can generate images, backgrounds, or even carousels to pair with your posts.

4. **Scheduling:** AI-powered tools like Buffer, Later, or Hootsuite can recommend optimal posting times.

Other prompts you can use to help with social media content include

- *Write 10 short tips about [topic].*

- *Generate 5 carousel post ideas for Instagram about [topic].*

- *Create 3 variations of this caption: one funny, one motivational, and one educational.*

- *Write a script for a 1-minute TikTok explaining [concept].*

- *Suggest 20 hashtags for posts about [topic].*

As a tip, remember that social media content should not only promote your product, but also educate, entertain, or inspire. AI helps you hit that balance by generating drafts you can polish into authentic posts.

EMAIL MARKETING

While social media gets the attention, email is still one of the most powerful tools for building a loyal audience. Unlike followers on social platforms, your email list is *yours*, a direct line to your audience, immune to algorithm changes. This usually means higher conversion rates compared to social media, the ability to build trust over time, and opportunities for automation.

In these cases, AI will help by drafting email newsletters, segmenting messages for different audiences, and writing catchy subject lines. It can even help you by creating automated sequences for onboarding or product launches. Suppose you are selling printable planners on Etsy and want to build an email list for repeat customers. This is an example of what this scenario would look like:

1. **Welcome sequence:** You can try the following prompt: *Write a 3-email welcome sequence for new subscribers who download a free printable planner. The tone should be friendly and encouraging.*

2. **Weekly newsletter:** The prompt, in this case, would be: *Draft a short, engaging email sharing one productivity tip and linking to my Etsy shop.*

3. **Product launch:** The following prompt could be used next: *Write a 5-email sequence to promote a new digital planner, including curiosity, urgency, and a final reminder email.*

4. **Subject line testing:** Finally, you can end the production by using: *Generate 10 catchy subject lines for an email about boosting productivity with daily planners.*

A few other prompt suggestions to help you in this process include

- *Write a welcome email for new subscribers interested in [topic].*

- *Create a 4-email nurture sequence that builds trust before a product launch.*

- *Draft a weekly newsletter template with 3 sections: tip of the week, product highlight, and motivational quote.*

- *Rewrite this email in a more playful tone suitable for young professionals.*

- *Summarize this blog post into a 200-word email teaser.*

There are also several tools you can use to automate this process. Within these, you will find free, freemium, and paid versions. In most cases, the free version will be enough to start out and for a small operation. Here are the tools you might want to explore:

- Mailchimp and ConvertKit for automation and list management.

- Beehiiv or Substack for newsletter-focused side hustles.

- ChatGPT/Claude for drafting and editing copy.

As you will see, the best part of AI is that it removes the friction from processes so that instead of having to wait weeks to draft a book, struggle to find time to brainstorm social media posts, or feel stuck at a blank email editor, you can start right now. With a few well-crafted prompts, you can create the building blocks of a digital side hustle in a single afternoon. With practice, you will discover that AI helps you do much more than just produce content; it will help you test, refine, and grow ideas faster than you thought possible.

BUILDING PASSIVE INCOME SYSTEMS

At this point, you have seen how AI can help you brainstorm ideas and create digital products. However, you should also know that you will find yourself in awe of what you are about to learn: These ideas can be turned into systems or processes that keep working for you even when you are not actively at your computer. This is the essence of passive income: building something once and allowing it to continue generating value over time.

On the other hand, it is essential to remember that passive income is rarely 100% passive. Most streams require setup, occasional maintenance, and updating. But compared to trading hours for wages, passive income allows you to scale. AI makes this even more attainable by reducing the time, cost, and expertise needed to create and maintain these systems.

Here are some of the most common passive income streams where AI can give you an edge:

- **Digital products**

- o *Ebooks, guides, templates, or printables that can be repeatedly sold.*
- o *AI helps generate content, design layouts, and create marketing materials.*

- **Content platforms**
  - o *Blogs, YouTube channels, or podcasts that earn through ads, sponsorships, or affiliate links.*
  - o *AI can draft blog posts, brainstorm video scripts, or edit podcast transcripts.*

- **Subscription models**
  - o *Membership sites, newsletters, or premium communities.*
  - o *AI supports regular content creation, engagement prompts, and audience insights.*

- **Affiliate marketing**
  - o *Recommending products with affiliate links and earning commissions.*
  - o *AI can write comparison articles, product reviews, and promotional emails.*

- **Print-on-demand**
  - o *Designing T-shirts, mugs, posters, or planners sold via platforms like Redbubble or Merch by Amazon.*
  - o *AI can brainstorm slogans, generate graphics, and create product descriptions.*

- **Stock assets**
  - o *Selling stock photos, music, or design assets.*
  - o *AI can help generate visuals, edit media, or brainstorm and create packs for marketplaces.*

While AI can do all these amazing things, it is normal to feel overwhelmed and even a little lost regarding which tool should be used for what purpose.

To help you with this, here is a quick reference table you can use to check out, test, and identify the best tool for your purpose and business.

Area / Activity / Skill	Tool
Content creation and writing	ChatGPT, Claude, Google Gemini
Visual design	Canva, DALL·E, Midjourney
Audio and video	Descript, ElevenLabs, Suno AI
Publishing and selling	Gumroad, Etsy, Amazon KDP, Teachable
Marketing automation	Mailchimp, ConvertKit, Beehiiv, Buffer
Print-on-demand	Printful, Redbubble, Merch by Amazon

These platforms pair seamlessly with AI outputs. For example, you can generate a blog post draft with ChatGPT, design its header image with Canva Magic Studio, and then schedule it on WordPress; all without hiring outside help.

WHAT AI CAN DO FOR YOU

Let's recap the benefits AI brings and the tasks it can handle across different income streams. Think of this as your "delegation menu":

- **Brainstorming:** Generate product ideas, niches, or blog topics.

- **Drafting:** Write articles, guides, or scripts for videos.

- **Editing:** Improve grammar, simplify language, or reformat content.

- **Designing:** Suggest visual concepts, generate templates, or draft cover art.

- **Marketing:** Write captions, create ad copy, or generate hashtags.

- **Customer engagement:** Draft FAQ responses, create onboarding emails, or write customer instructions.

- **System building:** Suggest automation workflows for email sequences, social posts, or publishing.

Each of these saves you time while also enabling you to focus on the strategic layer: choosing what to build, where to sell, and how to connect with your audience.

PROMPT TEMPLATES FOR PASSIVE INCOME SYSTEMS

Here is a prompt template library you can start using today. Enter your niche, product, or audience to tailor them.

- Digital products

    o *Outline a 20-page beginner's guide to [topic]. Suggest chapter titles and subtopics.*

    o *Write a 500-word draft for the introduction to my guide on [topic]. Make it friendly and motivational.*

    o *Create 10 printable planner templates for [audience]. Describe what each template should include.*

    o *Generate 5 cover design concepts for an ebook about [topic].*

    o *Write a product description for a digital course teaching [topic].*

- Blogs and content platforms

    o *Generate 20 blog post ideas for [topic] that could rank on search engines.*

    o *Write a blog post outline for the keyword [keyword]. Include H2 and H3 headings.*

    o *Create a YouTube script (3 minutes long) that explains [concept] to beginners.*

    o *Summarize this podcast transcript into a 700-word blog article.*

    o *Suggest 5 affiliate product review topics that fit within [niche].*

- Subscription models

    o *Draft the first three emails for a paid newsletter about [topic].*

    o *Suggest exclusive content ideas I could offer in a premium membership about [topic].*

    o *Generate 10 discussion prompts for a private community of [audience].*

- Write a monthly roundup email for subscribers about the latest news in [industry].
- Create a roadmap for a 12-month membership program teaching [skill].

- Affiliate marketing

  - Write a 1,000-word comparison article: [Product A] vs [Product B]. Include pros, cons, and recommendations.
  - Draft 5 promotional tweets highlighting the benefits of [product].
  - Generate email copy to promote an affiliate product for [audience].
  - Suggest 10 blog post topics where affiliate links for [product category] would naturally fit.
  - Write a product review in a casual, storytelling style about [product].

- Print-on-demand

  - Generate 20 T-shirt slogan ideas for [niche/audience].
  - Create 3 design concepts for a coffee mug about [theme].
  - Write product descriptions for 5 print-on-demand poster designs.
  - Brainstorm seasonal print-on-demand product ideas for [holiday/event].
  - Suggest 10 funny sticker ideas for [audience type].

- Stock assets
  - Generate 10 stock photo concepts for [industry or theme].
  - Suggest pack ideas for digital icons or graphics that small businesses could use.
  - Outline 5 niche sound effects people search for but rarely find.

- o *Create descriptions and keywords for uploading stock media to [platform].*
- o *Suggest 10 collections of digital backgrounds for presentations.*

EXAMPLE: A PASSIVE SYSTEM IN ACTION

Imagine you want to launch a passive income stream by selling printable templates on Etsy. This is the process you would follow:

1. Use AI to brainstorm ideas (*Generate 10 planner template ideas for busy parents*).

2. Pick 3 favorites: a meal planner, a homework tracker, and a bedtime routine chart.

3. Ask AI to draft the descriptions (*Write a product description for a printable meal planner, focusing on saving time and reducing stress*).

4. Create visuals in Canva using AI-generated design suggestions.

5. Upload them to Etsy and schedule AI-written social posts to promote your store.

6. Set up an email sequence (*Draft 3 emails for customers who download my free printable sample, encouraging them to check out the full planner pack*).

Within a week, you have moved from an idea to a system: a small shop that continues earning income as long as customers discover your listings.

Passive income is not magic; it is built on products, content, or assets that provide ongoing value. The challenge has always been the upfront work of creating and maintaining these systems. With AI, that becomes manageable and costless. You do not need to hire an expensive team or dedicate months of your life. Instead, you can use prompts, tools, and workflows to create repeatable systems.

BUILD A SMALL SIDE HUSTLE IN ONE WEEK

In the previous section, you were given six steps to start a business. But what if you were given more information so you could build a side hustle in one week? This is not about creating a million-dollar empire overnight. It is about creating a proof of concept: something small, scrappy, and real that shows you what is possible. Once your first small side hustle is live, you can refine it, expand it, or try another. The point is to start. Here is a walkthrough of what one week looks like.

DAY 1: BRAINSTORM IDEA

The first step is choosing something to build. Instead of waiting for a lightning-bolt inspiration, you will use AI to generate options quickly. The key is to focus on simple digital products such as guides, templates, checklists, printables, or short ebooks because they are easy to create, distribute, and sell. Some prompts you can use include:

- *Generate 15 digital product ideas for [audience].*

- *What pain points do new parents face that could be solved with a simple printable or guide?*

- *List 10 tiny side hustles I could start with less than $20.*

- *Suggest 5 product ideas related to [my hobby/interest].*

Case example:

Let's say you are into fitness. AI might suggest a 10-minute workout guide for beginners, a printable weekly meal planner for athletes, or a short ebook on stretching routines for desk workers.

Remember that you do not need to fall in love with the first idea. Just pick one that excites you enough to explore further.

DAY 2: RESEARCH YOUR AUDIENCE

Once you have an idea, the next step is figuring out who it's for and what they care about. AI makes this easier by simulating customer personas, analyzing niches, and suggesting messaging. Here are a few prompt examples:

- *Describe 3 types of people who would be most interested in a [product]. Include their motivations and challenges.*

- *What are the top 5 frustrations busy parents face when trying to cook healthy meals?*

- *Summarize common objections people might have about buying a [product].*

- *List online communities where people interested in [topic] hang out.*

Case example:

For the fitness meal planner, AI might identify your audience as busy professionals who want to eat better but lack time, college students on a budget, and parents seeking quick, family-friendly meals. Knowing this helps you shape your product and marketing.

DAY 3 AND 4: CREATE YOUR PRODUCT DRAFT WITH AI

Now it is time to build. Over 2 days, you will use AI to create a solid draft of your digital product. This could be an ebook (20–30 pages), a printable template or planner, a set of checklists, or a mini-course or email guide.

Step 1: Outline

Create a detailed outline for a [type of product] on [topic].

Step 2: Drafting content

Write a 1,000-word section for Chapter 2: Meal Prep Basics. Use a friendly and practical tone.

Generate 10 printable checklist ideas for organizing a weekly workout routine.

Step 3: Enhance with extras

Generate 20 motivational quotes about fitness to include as design elements.

Create a 7-day meal prep plan with shopping lists.

Case example:

By the end of day 4, your fitness meal planner could include a 20-page PDF with recipes, a printable weekly planner template, and a shopping checklist.

DAYS 5 AND 6: DESIGN AND SET UP YOUR STOREFRONT

A good product deserves a good presentation. These two days are for design and distribution.

Step 1: Design your product

1. Import your text

2. User premade templates for ebooks or planners.

3. Add AI-generated images.

Step 2: Set up your storefront

Choose a platform:

- Etsy: Great for printables and templates

- Gumroad: Simple for ebooks, guides, or digital packs

- Amazon KDP: For ebooks with a broader reach.

Upload your product, add pricing, and create a description. The following prompts can be used for this purpose:

- *Write a product description for Etsy that highlights the benefits of a printable meal planner.*

- *Draft 3 versions of a Gumroad sales page for my ebook about fitness and meal prep.*

Case example:

By the end of Day 6, you could have a polished product in a PDF format and a storefront page with attractive images, a compelling description, and a pricing set.

DAY 7: LAUNCH AND PROMOTE USING AI-GENERATED POSTS

Now, it's time to put your microhustle into the world. Launching does not have to mean paid ads or complicated campaigns. Start small, share with your network, and create a few AI-generated posts.

Step 1: Social media posts

- *Write an Instagram caption announcing my new meal prep guide. Make it motivational.*

- *Draft 5 post variations for X (formerly Twitter), introducing a digital product about fitness and meal planning.*

- *Write a LinkedIn post highlighting the value of meal prep for busy professionals.*

Step 2: Email announcement

- *Write a short, friendly email announcing my new digital guide. Include a link to purchase and a limited-time discount code.*

Step 3: Bonus promotion ideas

- Share in relevant Facebook or Reddit groups (respecting group rules).

- Offer a free sample page to encourage sign-ups.

- Partner with microinfluencers in your niche.

Case example: By the end of Day 7, you have shared your product on Instagram, posted on LinkedIn, emailed your small network, and listed it on Etsy. It may not make thousands overnight, but you have done something powerful: You have just launched a business.

This process works because it is broken down into daily steps, which prevent overwhelm. Each day builds momentum, and by the end of the week, you have a validated idea, a tangible product, a storefront ready for sales, and promotional content in circulation. The first version of your product might be simple, and that is okay. The point is to prove to yourself that you can build and ship. Once you see the first sale (or even the first piece of positive feedback), you will have the motivation to improve, expand, or try again with a new idea.

JOURNAL PROMPT

You have just walked through the process of building a tiny side hustle in a single week. Whether you followed every step or just imagined what it would be like for your idea, the point is this: you now see that starting is possible. You do not need a big budget, a team of professionals, or even months of preparation. With the right mindset, a willingness to experiment, and AI as your copilot, you can launch something real in days.

Before rushing into the next thing, it's worth pausing. Reflection turns activity into learning. IT helps you understand not just what you created, but what you discovered about yourself along the way. Here are your two prompts for this chapter:

What did you create this week that could generate value or income?

Take a moment to write down exactly what you produced. Did you draft an ebook? Create a printable planner? Set up an Etsy store? Even if it feels small or imperfect, give yourself credit for making something real. Then, go deeper: How does this product create value? Whose problem does it solve? What kind of person would benefit from it? This is more than about the money; it is about the usefulness and impact of what you have built. By focusing on value first, income will follow naturally.

What business task did AI make easier than you expected?

Think back on the week. Maybe AI helped brainstorm 20 product ideas in minutes, saving you hours of Googling. Maybe it wrote the first draft of your product description, something you have been dreading. Maybe it generated a social media post that made you think, *Wow, that actually sounds like me!* Capture the moment that surprised you the most, as this is where confidence grows.

By answering these two questions, you are doing more than journaling. The next time you face resistance, you will be able to look back at this page and remember, *I have done it before, I can do it again.* This reflection also helps you spot patterns, such as: Did you find brainstorming easy but design challenging? Did you enjoy writing content but struggle with setup? Did you feel energized by creating or more excited by promoting? These clues can guide your next project. Over time, you will discover the types of side hustles that best match your strengths and weaknesses.

THE BRIDGE FORWARD

So far, you have seen how AI can help you go from the idea to launch in just one week. That first win is powerful, and it proves that you can create something real with minimal resources and a lot of momentum. But once you have launched, the game changes. Running even the smallest side hustle comes with new challenges: keeping track of tasks, handling customer inquiries, writing updates, staying consistent with marketing, and finding time to actually grow the business.

This is where many people get stuck. They have built something exciting, but the day-to-day work feels overwhelming. The good news is that AI is not just useful for creation; it is also an incredible tool for streamlining operations. As we move on to the next part of this book, you will learn how AI can help you manage the behind-the-scenes of your business: automating repetitive tasks, simplifying communication, organizing workflows, and freeing up your time so you can focus on growth. Shall we start?

PART III:

AI IN THE WORKPLACE AND BUSINESS

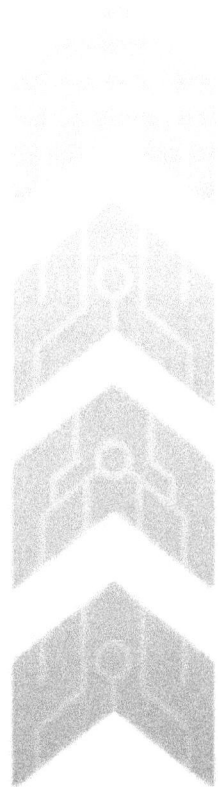

CHAPTER 7:

STREAMLINING SMALL BUSINESS OPERATION

CARLA, A SMALL BUSINESS OWNER, SITS AT HER KITCHEN table late at night, surrounded by receipts, her laptop, and a half-drunk cup of coffee. During the day, she manages her shop, chats with customers, updates her website, and monitors her social media. At night, she's her own book-keeper, HR department, and marketing team. Carla loves her business, but she is exhausted. She knows she is spending too much time on repetitive tasks such as sending invoice reminders, writing emails, and copying data from one system to another, but she does not know how else to keep everything on board.

Carla's story is hardly unique. Whether you are running a solo side hus-tle, a small company, or freelancing on the side, the real bottleneck is not usually a lack of good ideas. It is the weight of all the little tasks that eat away at your time, energy, and creativity. While large companies can afford entire operations teams to manage this load, solo entrepreneurs and small businesses don't have the luxury.

This is where AI steps in: With the right tools, you can suddenly have something that feels like a 24/7 operations team without all the payroll costs. AI can handle repetitive processes, draft communications, summarize meet-ings, analyze data, and even suggest better workflows. The result is not just more free time; it is the ability to shift your focus to higher-value activities: strategy, innovation, customer connection, and growth.

IMPORTANCE OF EFFICIENCY AND AI

In the past, success often went to the business with the largest budget or the loudest marketing. Today, however, agility is the real competitive edge. The businesses that thrive are those that can adapt quickly, reduce overhead, and make the most of limited resources. Efficiency is no longer a nice-to-have; it is the difference that will make your business survive.

Think of two Etsy shop owners selling similar products. One spends hours each week manually writing customer emails, organizing her lists, and planning social media posts. The other uses AI to automate replies, generate fresh content ideas, and keep track of inventory trends. Both have the same product quality, but the second shop owner has more energy to design new products and engage with customers, giving her an undeniable edge.

The beauty of AI is that it lowers the barrier to streamlining operations. You do not need to know how to code or invest in expensive enterprise software. Free or low-cost AI can help you

- draft routine emails in seconds.

- create chatbot scripts to answer FAQs on your website.

- summarize customer feedback into clear trends.

- turn messy spreadsheets into clear visual reports.

- draft company policies or meeting notes.

These may sound like small individual actions, but together, they can reclaim hours of your week. Imagine recovering 5–10 hours each week that you currently spend on repetitive work. What could you do with that time? Launch a new product? Take on more clients? Or simply reclaim a weekend?

SETTING EXPECTATIONS

It is important to note: AI won't magically run your business for you. It's not about handing everything over and walking away. Instead, think of it as delegation; teaching AI to manage routine, structured, and time-consuming tasks, while you focus on the creative and strategic work that only a human can do.

Throughout this chapter, you will explore practical ways to integrate AI into your operations, with examples, tool recommendations, and prompts you can adapt to your own context. You will learn how to

- automate repetitive tasks like emails and data entry.

- use AI for marketing and branding.

- improve internal processes like meeting notes and policy writing.

- audit your workflow to identify opportunities for automation.

By the end, you will see that you do not need a large staff or expensive systems to run an efficient, professional operation. You just need the right mindset and a willingness to let AI take some of the weight off your shoulders. Remember that efficiency is not about just doing things faster; it is also about creating the space to do the things that matter most.

AUTOMATING REPETITIVE TASKS

If there is one thing AI excels at, it is handling the small and repetitive tasks that sap your time and energy. These are the things that rarely move the needle on your business by themselves, yet are essential to keep things running. Think about the hours you have spent writing the same kind of email, copying and pasting data between spreadsheets, or answering the same customer questions over and over. It adds up.

The good news is that these are the exact types of tasks that AI can handle with remarkable efficiency. Let's focus on three key areas: emails, data entry/ analysis, and chatbots, and explore how you can automate them in your business without needing coding skills or expensive software.

AUTOMATING EMAILS

Emails are vital for most businesses, whether you are sending invoices, responding to customers, or staying in touch with your audience. But writing the same types of emails over and over is exhausting. AI makes it easy to generate drafts in seconds, which you can quickly personalize before sending. This will help you save time, keep communication consistent and professional, and reduce the risk of forgetting important details.

Tools to use:

- ChatGPT, Claude, or Gemini for drafting templates
- Mailchimp or ConvertKit for automation sequences
- Gmail add-ons like Flowrite that integrate AI directly into your in-box.

Prompt library:

1. Invoice reminders: *Draft a polite but firm email reminding a client their invoice is overdue by [X days]. Include payment options and a friendly thank-you.*

2. Customer support replies: *Write a professional response to a customer who received a damaged product. Apologize, explain the refund/return process, and offer a discount on their next purchase.*

3. Follow-ups: *Generate a short, friendly follow-up email to a client I spoke with last week about [topic]. Include a call to action to schedule a call.*

4. Event reminders: *Write a reminder email for attendees about tomorrow's webinar on [topic]. Include time, date, Zoom link, and a quick bullet-point agenda.*

5. Thank-you notes: *Create a warm thank-you email for a customer who just made their first purchase. Suggest related products and include a discount code.*

6. Newsletters: *Draft a 3-paragraph email for my weekly newsletter about [topic]. Include a top, a personal anecdote, and a link to my product.*

To help you save time, you can save your favorite drafts in a shared folder or an email automation tool. Over time, you will build a library of reusable templates that make responding to customers almost effortless.

DATA ENTRY AND ANALYSIS

Spreadsheets are a blessing and a curse: They are powerful for organizing information, but entering and analyzing data by hand is tedious. AI can help by writing formulas, cleaning up messy data, and even summarizing trends for you. This will help you save hours of manual entry and error-checking. It will also make analysis accessible to nonexperts and turn raw data into insights you can act on.

Tools to use:

- ChatGPT with code interpreter/advanced data analysis for complex spreadsheets.

- Excel with Copilot if you are using Microsoft 365 for built-in AI support.

- Google Sheets and add-ons like GPT for Sheets to directly integrate prompts.

Prompt library:

1. Creating formulas: *Write an Excel formula to create the average sales per week from column B* or *Create a formula that flags any cell in column C where the value is greater than 100.*

2. Cleaning data: *Here is a messy list of customer names and emails. Reformat it into a clean table with separate tables for first name, last name, and email address.*

3. Summarizing trends: *Summarize this sales data into a paragraph that highlights the best-performing products and months.*

4. Visualizing data: *Turn this table into a chart showing sales growth month by month. Suggest the best chat type to make the trend clear.*

5. Identifying outliers: *Analyze this data and identify any numbers that fall outside the normal range. Suggest possible reasons.*

Case Example

A small Etsy shop owner exported her monthly sales data into Google Sheets. Normally, she would spend hours calculating which products were performing best. Instead, she pasted the data into ChatGPT with the prompt: *Summarize this data into a report. Highlight my top 3 best-selling products, my slowest months, and suggest what I should promote in Q2.* In minutes, she had a clear summary and actionable insights, something that used to take her an entire afternoon.

CHATBOTS

One of the most time-consuming parts of a business is answering repetitive customer questions: "What are your hours?" "Do you ship internationally?" "How do I reset my password?" Chatbots allow you to automate these responses without needing to know how to code. This will free you from constant interruptions while giving customers instant support. Doing so will also allow you to provide instant responses to common questions, improve customer experience with fewer delays, and allow you to focus on higher-level conversations.

Tools to use:

- Tidio: Easy-to-use chatbot for small businesses

- ManyChat: Great for social messaging like Facebook and Instagram

- Landbot: Drag-and-drop builder for website chatbots.

Most of these have free versions and will require little to no coding. All you need to do is ask AI to help you write the scripts, which you will then paste into the builder.

How to create a chatbot script with AI

1. List the common questions. Think about the top 5–10 questions customers ask you.

2. Prompt AI to write drafts: *Write a friendly chatbot script answering these FAQs [list your questions]. Make the responses concise and polite.*

3. Add personality: *Rewrite these chatbot responses in a playful, upbeat tone that matches my brand voice.*

4. Create branching options: *Write a chatbot flow for a small online clothing store. Include branches for sizing, shipping, returns, and discounts.*

Try to keep the responses short, between 1–3 sentences, to ensure the AI does not hallucinate. It is also good practice to give the customer the option to contact a human if needed. Finally, remember to test the chatbot yourself before launching; your bot's credibility might depend on it!

ONE WEEK WITH AI IN COMMAND

Here is what a typical week for an online business would look like:
- On Monday, AI drafts invoices that go out automatically.
- On Tuesday, AI cleans and summarizes last week's sales data, giving you insights for the next campaign.
- On Wednesday, your chatbot handles 70% of customer questions, freeing you to work on new projects.
- By Friday, you have saved 6–8 hours of repetitive work. This time, you can now spend on growth activities or simply reclaim for yourself.

At the same time, remember that you do not need to automate everything all at once. Start with one area: emails, spreadsheets, or chatbots, and see how much time you save. Once you feel the relief, you will naturally start looking for other ways to delegate the busywork. That is the true competitive advantage.

AI FOR MARKETING AND BRANDING

If you talk to most business owners, freelancers, or creators, you will often hear the same sigh when the subject of *marketing* comes up. Most say, "I know I need to do it, but I don't have the time." How many times have people been in this same situation because there was no space for creativity in their busy schedules?

Take Jasmine, who runs a small online jewelry shop. She adores designing pieces and chatting with her loyal customers, but every week she dreads sitting down to write Instagram captions and plan posts. "I spend more time staring at a blank screen than actually posting anything," she confessed. Or Marco, a personal trainer who loves working with clients in the gym but feels overwhelmed by the pressure to produce TikTok videos to stay visible. "I am not a content creator," he says, "I just want to train people."

Stories like Jasmine's and Marco's are everywhere. Marketing and branding take time, consistency, and creativity, all things which most entrepreneurs are short on. The irony is that without marketing, even the best product or service can fade into obscurity.

With the right tools, AI can feel like hiring a marketing assistant who is available 24/7, does not get tired, and always has fresh ideas. From drafting captions to designing visuals, planning campaigns, and helping you find brand voice, AI can dramatically lighten the load while still leaving room for your unique personality to shine.

SOCIAL MEDIA POSTS

Have you ever felt as if social media is a treadmill from which you cannot step off? There is constant pressure to post consistently, stay on trend, and balance creativity with strategy. AI can help you create posts, generate hashtags, and even design visuals that align with your brand.

Tools to use:

- Canva's Magic Write and Magic Design: Type in your idea, and it generates captions and matching graphics.

- DALL·E or Midjourney: Create original visuals to pair with your posts.

- ChatGPT or Claude: Draft multiple caption variations so you can pick the one that feels the most natural.

Prompt library:

- *Write 3 Instagram captions for a handmade jewelry business. One should be playful, one elegant, and one focused on craftsmanship.*

- *Generate 5 trending hashtags for a post about [topic]. Rank them by popularity.*

- *Suggest 3 TikTok video ideas for a personal trainer who wants to teach quick fitness tips in under 60 seconds.*

Imagine Jasmine again: Instead of spending 2 hours struggling with captions, she spends 10 minutes feeding prompts into ChatGPT, picks the best option, and pastes it into Canva, where the tool automatically creates a matching graphic. Suddenly, social media feels less like a chore and more like a streamlined part of her workflow.

CONTENT PLANNING

Posting one-off updates is not enough to build a strong online presence. The real engagement begins when you tell a consistent story over time. But sitting down to create a three-month content calendar can feel like too much: What do you post, when, and in what format? This is another area where AI can help you. By feeding it details about your business, goals, and audience, you can generate an entire plan of themes, post types, and schedules.

Sample prompt:

Create a 4-week content calendar for a small bakery. Include 3 Instagram posts per week, 1 TikTok idea, and 1 blog post. Mix in promotional, educational, and behind-the-scenes content.

The output might be something that looks like this:

- Week 1: Post a photo of your best-selling pastry with a story about its origin.

- Week 2: Share a behind-the-scenes video of dough being prepared.

- Week 3: Educational post about why sourdough is healthier.

- Week 4: A limited-time offer for loyal customers.

Tools like Notion AI, Trello with add-ons, or ClickUp AI make it easy to generate the plan and keep track of everything organized.

BRAND VOICE DEVELOPMENT

One of the most overlooked aspects of marketing is consistency in tone. Do you want to sound playful and casual or authoritative and professional? Too often, small business owners switch tones without even realizing it, leaving customers confused about what their brand stands for. Here, AI can act as a coach, helping you test and refine your voice. By experimenting with prompts, you can see how different styles resonate and decide what feels authentic.

Prompt library:

- *Write 3 versions of a product description for my eco-friendly bottle: 1 in a professional tone, 1 playful, and 1 poetic.*

- *Rewrite this paragraph in a tone that matches Apple's marketing style.*

- *Based on this sample text, identify my brand voice and suggest 3 ways to make it more consistent.*

Once you find your voice, you can tell AI, *From now on, write in a friendly, approachable, slightly humorous tone that makes the reader feel like we are chatting over coffee.* And just like that, every draft it produces will align more closely with your brand personality.

Let's take a minute to circle back to Marco, the personal trainer. Before AI, he dreaded marketing. Now he spends Monday morning asking ChatGPT to draft a week's worth of Instagram captions, complete with hashtags. He plugs them into Canva, which automatically designs matching graphics. Then, he uses Notion AI to generate a 30-day calendar of TikTok video ideas.

Instead of piecing together random posts, he now has a consistent plan, a recognizable voice, and visuals that fit his brand. Customers notice too: His engagement grows, and he starts booking more sessions. He did not suddenly become a marketing genius; he leveraged AI to make the process manageable.

REFLECTION

Marketing does not have to feel like shooting into the void or constantly running to catch up. With AI, you can approach it strategically, consistently, and without burning yourself out. And do not worry about keeping our own voice. AI handles the heavy lifting while you add the human spark that connects with your audience.

In the past, only large companies with marketing teams could achieve this level of polish and consistency. Now, even solo entrepreneurs can project the same level of professionalism without the stress.

AI will not replace your creativity, but it will give you the time, tools, and structure to let that creativity shine. When your marketing reflects your true brand voice, your customers don't just buy your product; they connect with your story.

IMPROVING INTERNAL PROCESSES

When some people think about AI in business, they often picture flashy marketing campaigns or futuristic customer service robots. However, some of the most powerful and underrated applications are behind the scenes, in the internal processes that keep a business running smoothly.

These are the activities few people see, but every business depends on: meetings, policy-making, and decision-making. They are also the activities that often drain time, create frustration, and leave people feeling overwhelmed. Ask any entrepreneur about meetings that run too long, or policies written in legalese that no one follows, or decisions made with too little information, and you will see why these areas are ripe for improvement.

AI can act like a quiet operations assistant, smoothing the edges of these processes. It does not eliminate them, but it makes them sharper, faster, and more useful. Here is how AI can help with meeting summaries, policy drafting, and decision support.

MEETINGS THAT DON'T WASTE TIME

Most business owners have a love–hate relationship with meetings. They are necessary for alignment, but they often feel like time sinks. People talk in circles, the key points get lost, and a week later, no one remembers action items. AI can help you with this by turning messy conversations into clear summaries and action plans.

How it works:

- Record your meeting on Zoom, Google Meet, or even your phone.

- Use a transcription service like Otter.ai, Fireflies.ai, or Norra to get a text version of everything said.

- Paste the transcript into an AI tool like ChatGPT or Claude and prompt it to distill the content.

Prompt library:

- *Summarize this meeting in 5 bullet points. Include decisions made, open questions, and next steps.*

- *From this transcript, extract a to-do list with deadlines and who is responsible.*

Real-Life Story

Daniel runs a small digital agency. He used to leave meetings feeling like nothing had been accomplished. Now, every meeting is recorded, transcribed, and run through AI. Within minutes, he has a one-page summary with three action items, two client concerns, and one idea to explore. His team knows what to do next, and meetings that used to take 90 minutes now take 45. In this case, AI also improves accountability, since action items are clearly written down, leaving less room for confusion or forgotten tasks.

POLICY DRAFTING WITHOUT HEADACHE

Policies are another idea where small businesses often struggle. You know how you need them: guidelines for how employees should act, how customer complaints are handled, or how refunds are processed. However, writing them from scratch feels intimidating. You are not a lawyer, and you do not want to sound like one, either.

How to use AI with policy drafting:

- Start with your vision and mission: *Our bakery values community sustainability and kindness,* for example.

- Prompt AI to generate a draft: *Write a refund policy for a bakery that values community and kindness. Keep it short and easy for customers to understand.* Or, *Create an employee code of conduct for a small design studio that emphasizes collaboration, respect, and creativity.*

- Refine and customize. Don't just copy and paste. Read carefully the document that is created, check, and adjust for legal accuracy. AI provides a draft, but you remain the decision-maker.

Example Output

Refund policy:

We want you to love every purchase. If you're not satisfied, you can return any bakery item within 7 days for a full refund or exchange. Please bring your receipt and let us know how we can make things right. Because our products are perishable, we cannot accept returns after 7 days.

Tools to try:

- ChatGPT or Claude for drafting.

- Notion AI to keep policies organized and accessible to your team.

- Canva Docs with Magic Write if you want visually polished policy documents.

A note of caution:

AI is not a lawyer. Always double-check for compliance with local regulations and, if necessary, have a professional review important policies. You can use AI as the starting point that needs refinement, but it is essential to review the final result.

DECISION SUPPORT

Decision-making is at the heart of every business. What products should you launch next? Should you hire another team member? Is it worth investing in a new tool or advertising channel? The challenge is that decisions often involve incomplete information, personal biases, and numerous variables. AI can't tell you the "right" decision, but it can act like a sounding board. With it, you will have a clearer vision of the pros and cons, analyze the context, and suggest perspectives you might not have considered.

HOW TO USE AI FOR DECISION SUPPORT

1. Clearly frame the question:

   o *Should I expand my product line to include gluten-free products?*

   o *Is it worth paying for a professional social media manager, or should I continue doing it myself?*

2. Provide context. The more details you include, the better the output:

o   *I run a small café in a city with a growing gluten-free community. Right now, 10% of customers ask about gluten-free options. Offering them would require new ingredients and staff
training.*

3.   Prompt for balanced analysis:

o   *List 5 pros and cons of adding gluten-free items to my menu.
Consider customer demand, costs, and brand reputation.*

o   *Suggest three alternative solutions to hiring a social media manager that could still reduce my workload.*

EXAMPLE OUTPUT

Pros of adding gluten-free items:

- Expands customer base.

- Shows inclusivity and responsiveness to trends.

- Potential premium pricing.

Cons:

- Higher ingredient costs.

- Risk of cross-contamination without training.

- Adds complexity to the menu.

Alternative: Partner with a local gluten-free bakery instead of making the
products in-house.

Suddenly, the decision feels less overwhelming. You have a clear picture
of trade-offs and new ideas you might not have thought of.

TOOLS THAT HELP

- Perplexity AI for great research and gathering context across the web.

- ChatGPT or Claude for structured analysis and brainstorming scenarios.
- Notion AI or Trello AI add-ons to keep decision notes organized for future reference.

THE AI ADVANTAGE

To understand how this all fits together, think about Sarah, who runs a five-person marketing agency. Before integrating AI, team meetings were long, and no one remembered the next steps. Policies were scattered across random Google Docs, half-written and outdated. Finally, business decisions felt like gut calls, with little time to weigh alternatives.

Now, with AI, every meeting ends with a one-page summary and clear action items. Policies are drafted in plain language, aligned with company values, and updated regularly. Lastly, decisions are approached with structured pro and con lists and creative alternative strategies.

Sarah did not eliminate meetings, policies, or decision-making. What she did was reduce the amount of time that she needed to dedicate to each topic by using AI logic. This led to more available time, better outcomes, and less stress. Her team notices the differences, too. They feel more aligned, less confused, and more confident about the company's direction.

Improving internal processes may not sound glamorous compared to designing logos or creating social media campaigns. But in reality, these are the links that hold your business together. Clear meeting notes prevent confusion, straightforward policies protect your values and customers, and balanced decision-making reduces costly mistakes.

AI will not take over leadership: It will not decide your strategy or enforce your culture. But it can give you the structure and clarity to make better use of your time and energy. Often, that is the difference between a business that feels chaotic and one that feels sustainable.

USE AI TO EDIT YOUR WORKFLOW

Every business has bottlenecks, those tasks that eat up too much time, feel repetitive, or drain energy without adding much value. Left unchecked, these tasks pile up and turn even the most passionate business owner into a frazzled multitasker.

But what if you could step back, shine a light on those bottlenecks, and systematically tackle them? That is what auditing your workflow with AI allows you to do. It is like having a consultant look at how your business runs and suggest smarter and faster ways to operate. The only difference is that instead of hiring an expensive consultant, you can use tools you already have.

WHY AUDIT YOUR WORKFLOW?

Raj runs a small marketing agency. He is proud of his creative work, but most days, he feels like he is drowning. Between scheduling meetings, chasing invoices, drafting customer reports, and posting on social media, he barely has time for the creative brainstorming he loves. When a friend suggested auditing his workflow, he rolled his eyes and claimed that he did not have the time to analyze how he worked. However, when he finally tried it with AI's help, he realized that 60% of his week was spent on tasks he could delegate or automate. By making those changes, he freed up an entire day each week.

Here is how you can do the same.

Step 1: List Your Time-Consuming Tasks

The first step is simple: Grab a pen and paper or open a document on your computer and list 5–10 tasks you do regularly that take too much time. These are often things you dread or put off.

Examples might include

- answering repetitive emails,

- entering data into spreadsheets,

- managing your social media posts,

- creating invoices and sending reminders,

- drafting reports or proposals,

- scheduling appointments, and

- following up with clients.

The point here is not to be exhaustive, but to identify the top culprits that drain your energy.

Step 2: Use AI to Draft Solutions

Once you have your list, it is time to bring AI into the picture. Instead of staring at the list and feeling overwhelmed, you can ask the AI, *What is the best way to handle this?* Other prompts you can use include

- *I spend 5 hours a week answering repetitive customer emails. Suggest 3 ways I can use AI to automate or speed up this task.*

- *Here are 10 tasks I do every week: [list tasks]. Which of these could AI handle most effectively, and what tools should I use?*

- *Draft a step-by-step plan for how I can use AI to reduce the time I spend on invoicing and payment reminders.*

Step 3: Choose 3 Tasks to Redesign

You do not need to overhaul your entire workflow at once. Start with three tasks that would make the biggest difference if you automated or streamlined. These should be

- frequent (you do them weekly or daily),

- time-consuming (they take at least 30 minutes), and

- structured (there is a clear process AI can follow).

Here are three examples:

- Task 1: Drafting weekly reports

- Task 2: Scheduling meetings and reminders

- Task 3: Following up on unpaid invoices

Then, prompt AI to give you a playbook for each.
Prompt library:

- *Write a workflow for creating weekly client reports using AI. Include what data I should input and how AI can format it.*

- *Suggest an automation plan for scheduling meetings using free AI tools.*

- *Write 3 polite invoice reminder templates I can use in my email automation tool.*

Step 4: Create a Workflow Map

Now comes the fun part: turning your messy list of tasks into a visual workflow map. This helps you see how tasks connect and where AI can slot in. You do not need fancy software to do this. A whiteboard, notebook, or free tools like Micro, Lucidchart, or even Canva work perfectly.

How to Build Your Map

1. Write each task in a box. Example: Client sends request → I write response → I send report.

2. Mark the steps AI can handle. Ex: AI drafts response → I review → I send report.

3. Add arrows to show flow. Seeing the process visually helps you identify redundancies or areas for automation.

Example

Imagine your old invoicing process looks like this:

- Manually create invoice → email client reminder → track payment in spreadsheet

New workflow (with AI):

- AI tool generates invoice from template → Automated email drafted by AI reminds the client → AI updates spreadsheet and summarizes overdue accounts weekly

Suddenly, a three-step time drain becomes a streamlined system.

TOOLS THAT CAN HELP

- ChatGPT and Claude for brainstorming automation solutions and writing scripts and templates.

- Zapier and Make to connect apps and automate repetitive actions.

- Otter.ai and Fireflies.ai for meeting transcriptions and summaries.

- Canva + Magic Write for quick social media creation.

- Google Sheets with GPT add-ons for data entry and analysis automation.

Remember that the goal is not to master every tool, but to identify where AI naturally fits into your existing systems.

When Raj finally mapped his workflow, he was stunned. What felt like "the way things had to be" turned out to be riddled with inefficiencies. His invoicing was manual. His clients' reports were written from scratch. His meetings went undocumented. With AI's help, he redesigned three workflows, and by the end of the month, he had saved 20 hours.

Automating your workflow is about taking control of your time and choosing smarter ways to work. By listing your top tasks, asking AI for solutions, and visualizing the new process, you give yourself the gift of clarity.

TRY IT NOW

Here is a quick exercise:

1. Write down 5–10 tasks you do every week.

2. Circle 3 that frustrate you the most.

3. Ask AI: *What is the best way to automate or streamline these 3 tasks using free tools?*

4. Sketch a workflow map that shows where AI fits in.

You might be surprised by how much lighter your workload feels once you see it on paper.

Auditing your workflow is more than an efficiency exercise; it is a mind-set shift. It teaches you to stop accepting the way it has always been and start asking, *Is there a smarter way?* AI is your partner in that process. You do not need to transform your business overnight. Just start with 3 tasks and, once you see the results, you will naturally begin to rethink others. Over time, your business becomes leaner, smoother, and more resilient.

JOURNAL PROMPT

You now know how AI can step into the background of your business as a reliable operations partner. From automating emails and creating chatbots to drafting policies and auditing workflows, you have seen how even the smallest adjustments can give you back hours of your week.

The real transformation will happen in what you do and how you think. Using AI is not only about efficiency, but also about learning to let go of work that does not need your constant attention. That can feel liberating, but it can also feel strange at first. If you have always prided yourself on doing everything on your own. The idea of handing off tasks can trigger a little resistance.

For this chapter, your journal prompts will be:

Which tasks felt easiest to hand off to AI?

Think back to the examples in this chapter, or even the tasks you have already tested. Write about what felt surprisingly simple to let go of. Why do you think that was? Was it because the task was repetitive and boring? Because AI's draft felt good enough without editing? Because you didn't need to do it yourself after all? Capturing these observations helps you build trust in the process. Once you recognize the tasks that are safe to hand off, you will find it easier to identify the next ones.

What is one process you will fully automate in the next 30 days?

Now it is time to get specific. Out of all the tasks you have explored, which one makes the most sense to automate right now? Pick just one. Maybe it is automating client follow-ups, creating weekly reports, or scheduling social media posts. Write down what this automation would look like in practice. What tool will you use? What prompts will you rely on? What small role will you still play? And most importantly, how much time will you save?

Auditing workflows and automating tasks is not just a technical exercise. It is about clarity, confidence, and reclaiming control over your workday.

By journaling about what was easy to hand off and what you will automate next, you are building a road map for yourself, one that keeps your business efficient without overwhelming you. Remember: Every hour you reclaim is energy you can reinvest in growth, creativity, or simply taking a breath. The small wins add up faster than you might think.

So far, we have looked at how AI can help you as a solo operator: streamlining how you handle tasks, workflows, and operations. But what happens when you are not working alone? Most businesses, whether big or small, depend on teamwork. Regardless of whom you are collaborating with, managing a small staff, or working with freelancers around the world, the way you coordinate and communicate can make or break your success.

That is where we are headed next: In Chapter 8, you will learn how to use AI to coordinate projects, improve communication, reduce misunderstandings, and even strengthen team culture. Just as AI can take busy work off your plate, it can also act as a connector between people, helping teams run more smoothly, stay aligned, and achieve more together.

AI FOR TEAMS AND COLLABORATION

LENA RUNS A SMALL DESIGN AGENCY WITH FIVE EMPLOY-ees. On paper, the team is strong: talented designers, an eager intern, and a dependable project manager. However, most mornings, Lena dreads opening her email inbox. It is a flood of half-finished threads on several subjects. She spends more time trying to piece together what her team is doing than actually doing the work she loves.

This is the reality for many teams, whether you are in a start-up, nonprofit, or even a community group. Collaboration is supposed to make things easier, but often it just creates chaos: endless emails, scattered notes, and meetings that drain more energy than they give. The bigger the team, the bigger the mess.

Now, imagine Lena had a digital facilitator that never got tired, always remembered the details, and could instantly summarize what had been said or agreed on. A tool that could generate meeting agendas, recap discussions, remind everyone of their tasks, and even translate messages for a global client. That is what AI can be: not a replacement for human teamwork, but a quiet, steady presence that brings order to the noise.

Most teams underestimate the amount of time wasted on disorganized collaboration. A report from McKinsey once estimated that employees spend nearly 20% of their workweek just searching for information (Probstein, 2019). Add to that the hours lost in repetitive meetings or correcting miscommunications, and the cost of chaos becomes painfully clear.

The problem is not that teams do not want to work well together. The tools they have traditionally used (email, shared drives, and chat apps) were not designed for the sheer speed and volume of modern work. That is why AI is so powerful: It acts like a facilitator who is neutral and tireless.

AI AS A DIGITAL FACILITATOR

So what does this look like in practice? Here are a few examples:

- **Organizing:** AI can generate an agenda before a meeting, pulling in key topics from recent conversations and meetings. No more showing up to a Zoom call wondering what's on the table.

- **Summarizing:** Instead of everyone frantically scribbling notes, AI can produce a concise recap with clear action points. The conversation stays human, and the record-keeping is automatic.

- **Coordinating:** AI can nudge team members when deadlines are approaching, suggest task assignments based on workload, and keep everyone aligned without micromanagement.

Think of AI as the team member who never loses track of the thread. While humans bring creativity, empathy, and judgment, AI quietly ensures nothing falls through the cracks.

WHY THIS MATTERS NOW

Remote and hybrid work have amplified the need for better coordination. When teams are not in the same room, the risk of miscommunication skyrockets. AI helps close these gaps and bring people together. For example, a global nonprofit might use AI-powered translation so volunteers in Kenya, Germany, and Brazil can all contribute to the same project without language becoming a barrier. A marketing team spread across three time zones might rely on AI to provide a single source of truth after brainstorming sessions, ensuring no good idea is forgotten.

Collaboration without chaos is about amplifying the human element, not reducing it. When the logistics are handled, people are free to focus on what humans do best: solving problems, building relationships, and creating new ideas.

This chapter is about reimagining teamwork with AI as a digital collaborator, not as a boss or a replacement. You will learn how AI can

- help your team communicate better with agendas, summaries, and even translations.

- act as a brainstorming partner who sparks ideas and then narrows them into plans.

- support project management by suggesting timelines, assigning roles, and tracking progress.

- facilitate collaborative exercises where human creativity and AI efficiency work side by side.

By the end, you will see that collaboration does not have to be messy or draining. With AI, it can be structured, creative, and even joyful.

AI FOR TEAM COLLABORATION

If collaboration is the engine of teamwork, then communication is the fuel. When you ask teams what slows them down, miscommunication is almost always near the top of the list. Meetings that ramble without direction. Notes that vanish into inboxes. Language barriers leave teammates nodding politely but silently confused.

AI offers relief. Not because it takes over the conversations (human connection still matters most), but because it can handle the parts of communication that are most prone to mess: planning, recording, and translating. Let's break this down into three areas: meeting agendas, summaries and recaps, and language translation. Each shows how AI can take your existing team communication and make it sharper, faster, and more inclusive.

MEETING AGENDAS

Everyone has been in "that meeting" where, within five minutes, no one knows why they are there. The conversations drift from one topic to another, and an hour later, everyone leaves, wondering what (if anything) was accomplished. The culprit is the lack of clear agendas. While these agendas may sound boring, they are the skeleton of a productive meeting. They set expectations, keep discussions focused, and ensure time is used well. The problem is that busy teams rarely have the time to draft them, and that is where AI can help.

AI creates agendas by being fed a bit of context, like the meeting's purpose, recent emails or chat logs, and the expected attendees, allowing you to create a structured agenda in a few minutes. Here is a prompt example you can use: *Draft a 45-minute agenda for our weekly marketing sync. Topics include campaign performance, upcoming launches, and brainstorming new social media ideas. Include time blocks and who should lead each section.*

A sample output could be

- **Welcome and Objectives (5 min)** – Team Lead

- **Campaign Performance Review (10 min)** – Analytics Manager

- **Upcoming Launches (15 min)** – Product Marketing Lead

- **Brainstorm: Social Media Ideas (10 min)** – Open Discussion

- **Action Items and Next Steps (5 min)** – Project Manager

This way, instead of starting from scratch, you have a professional framework you can tweak as needed. Tools that might help you accomplish this include

- ChatGPT, Claude, or Gemini for quick text-based agendas.

- Otter.ai or Fireflies.ai, which allow the combination of transcription and agenda features.

- Notion AU to integrate agendas into collaborative documents that your team can access.

The effect is subtle but powerful, where people show up knowing the meeting's goals and structure. This allows decisions to get made faster, discussions to run more smoothly, and everyone to leave knowing their time was respected.

SUMMARIES AND RECAPS

If agendas prevent wasted time at the start of meetings, summaries prevent wasted effort after. Too often, teams leave meetings with vague memories of good conversations, but no one remembers the exact decisions or next steps. A week later, someone asks, "Wait, did we really talk about this?" and the cycle repeats. AI solves this by acting as the team's note-taker, only faster, sharper, and always consistent.

The first thing you need to do is record the meeting. Then you upload it to AI software like Otter.ai or Notta and prompt the tool for summaries and action items. Prompts such as, *Summarize this transcript in a one-page recap. Highlight key decisions, unresolved questions, and action items assigned to individuals*, or *create a bulleted to-do list with deadlines based on this meeting transcript.*

The AI tool might give you something like this:

- **Decisions Made:** Launch date for beta moved to June 15. Budget approved for design contractor.

- **Unresolved Questions:** Need confirmation from legal about privacy policy.

- **Action items**

    o   Alex: Draft press release by May 30.

    o   Priya: Coordinate contractor onboarding by June 1.

    o   James: Follow up with the legal team.

Consider Martha, who runs a small nonprofit with volunteers scattered across three states. Weekly calls were full of passion and ideas, but no one wrote things down. Two months later, projects lagged because no one was sure who agreed to what. After adopting AI summaries, Martha emailed a one-page recap after each call. Volunteers stopped forgetting tasks, and projects moved faster without her personally having to spend hours writing notes.

Tools that can help you streamline this include most of the tools that you have seen so far. They are

- Otter.ai or Fireflies.ai for live transcription and AI-generated summaries. Zoom also has the same feature for video calls.

- Notion AI is great for creating shared knowledge bases with recaps stored by project.

- ChatGPT or Claude can be used when you want to customize the summary format.

LANGUAGE TRANSLATION

Global teams are no longer rare; they are now the norm. Freelancers in Manila work with clients in Toronto. Start-ups in Berlin collaborate with engineers in São Paulo. While this global reach is exciting, language differences can create subtle barriers. Even when everyone speaks "business English," nuances and confidence can be lost. AI-powered translation tools make collaboration more inclusive by breaking down those barriers. They can translate documents, live chats, or even spoken conversations in real time.

Tools that can help you explore this potential include

- DeepL is widely praised for natural-sounding translations, especially for European languages.

- Google Translate is a more versatile and language-inclusive system.

- Microsoft Translator integrates with Teams for real-time captions.

- ChatGPT (with transcription prompts) can be used for quick written translations with tone adjustments.

Prompts you can use to achieve this might read something like this:

- *Translate this project update into Spanish, keeping the tone professional but friendly.*

- *Summarize this meeting transcript in French, focusing only on the action items.*

- *Rewrite this customer support email in Japanese with polite phrasing appropriate for business.*

Ana is based in Mexico and works with a US-based marketing agency. In early meetings, she felt hesitant to share ideas, worried her English was not polished enough. When the team started using Microsoft Teams' live translation, everything changed. Ana could speak in Spanish, and her colleagues instantly read English captions. For the first time, her creativity flowed freely in meetings, and the team benefited from ideas they might have missed before.

Keep in mind that it is not a matter of replacing team communication, but rather enhancing it. Meetings are sharper when expectations are set and outcomes stick when summaries capture decisions. The result is teams that feel stronger when language stops being a barrier. Imagine a week in a small business using these tools:

- Monday: The weekly planning meeting runs on time because AI drafted the agenda.

- Tuesday: A client call ends, and within 15 minutes, AI delivers a recap with clear action items.

- Wednesday: A global brainstorming session flows naturally, with real-time translation bridging languages.

- By Friday, the team feels aligned, energized, and confident, not buried under messy notes or missed details.

PROMPTS TO TRY THIS WEEK

- *Draft a 60-minute agenda for a cross-functional team meeting about launching a new product. Include time for updates, brainstorming, and assigning next steps.*

- *Summarize this transcript into 5 bullet points of key takeaways. Include 3 action items with names attached.*

- *Translate this project plan into Portuguese with a professional but motivating tone.*

- *Rewrite these meeting notes in a format suitable for sharing with executives. Keep it high-level and concise.*

- *Draft a client call summary that includes project updates, feedback, and next steps. Keep the tone professional and reassuring.*

Communication breakdowns are one of the top productivity killers, but they do not have to be inevitable. By letting AI handle agendas, recaps, and translations, your team gets back to what matters the most: time to think, space to create, and the confidence that nothing important will slip through the cracks. Collaboration without chaos begins one agenda, one summary, and one translation at a time.

BRAINSTORMING AND IDEATION

If you have ever sat in front of a blank page, you know how heavy silence can feel. The blinking cursor taunts you and whispers, *What if you do not have any good ideas?* Teams experience this tool. A group gathers in a meeting room or on Zoom for a brainstorming session. The goal is to dream big, but what actually happens? Half the group stays quiet, and a few voices dominate. Someone throws out a suggestion that gets politely ignored, and another idea sparks an endless debate. By the end, everyone feels drained, and the whiteboard looks emptier than expected.

This is where AI becomes your cocreator, helping you unlock your creativity. Once again, it is important to emphasize that AI will not replace you; your human qualities still need to be added to the final output. However, AI can take the pressure off by generating starter ideas, exploring wild possibilities, and helping teams narrow down to realistic options.

COCREATION WITH AI

Eva is a content strategist for a restaurant chain. She often leads brainstorming sessions for campaign ideas. Before AI, she would prep for hours, collecting inspiration so she would not freeze in front of her colleagues. Now, she enters the session with a list of AI-generated prompts in her pocket. If the team goes quiet, she can say, "Here are the five concepts AI suggested; we can build on them." Suddenly, the pressure is not on one person to be brilliant. The AI provides raw material, and the team shapes it into something meaningful.

A few prompts you can try for this purpose include

- *Generate 20 ideas for a social media campaign for a coffee shop that wants to appeal to young professionals.*

- *Suggest unusual themes for a podcast about entrepreneurship. It should be something that stands out from typical business shows.*

- *Come up with 10 product names for a new line of eco-friendly candles. Include playful, elegant, and minimalist options.*

- *Draft 3 headlines for an article about remote work, each in a different tone: casual, formal, and humorous.*

Notice how none of these outputs are "final." They are sparks that will get the team started on debates, refining, and deciding, since they will be starting from abundance instead of scarcity.

DIVERGENT THINKING: GOING WIDE

Brainstorming works best when you allow wild, even silly, ideas to surface. This is called divergent thinking: exploring as many possibilities as possible without judgment. Humans often hold back here because they do not want to look foolish. AI, on the other hand, has no ego.

Imagine you are an events company trying to brainstorm fundraising activities. Left to themselves, they might suggest the usual suspects: gala dinner, silent auction, bake sale. When they ask AI for ideas, they get a wild list: "A dog fashion show, a 24-hour online dance marathon, and a community art mural project." Some are unrealistic, but others spark laughter and then genuine interest. The "dog fashion show" idea? Too silly for the company's brand. But the community art mural project? Suddenly, that feels fresh, aligned with their mission, and full of potential.

Try some of these prompts on your computer:

- *Suggest 15 unusual fundraising events for a local nonprofit. Make them creative, surprising, and fun.*

- *Give me 20 different plot twists for a mystery novel where the main character is a chef.*

- *List 10 outrageous marketing stunts a small sneaker company could try. Don't worry if they are impractical.*

The point is not to use these ideas, but to break the mental ruts that keep you circling the same old solutions.

CONVERGENT THINKING: NARROWING DOWN

Of course, ideas are not enough. At some point, you need to pick the ones worth pursuing. This is where convergent thinking comes in, narrowing the field, analyzing opinions, and deciding what is realistic. AI can help here, too, by organizing and evaluating. Instead of staring at a messy whiteboard of ideas, you can ask AI to cluster them into themes, weigh pros and cons, or even score them against your goals.

A start-up is brainstorming features for a new wellness app. They end up with 30 sticky notes of ideas ranging from guided meditation to AI-powered dream analysis. Instead of aimlessly debating, they paste the list into ChatGPT and ask: *Group these 30 app feature ideas into 5 categories. Then suggest which 3 ideas are most feasible for a 1st release, considering cost, time, and user appeal.* The team does not blindly follow the AI's advice, but it gives them a clear framework to discuss.

Here are 3 other prompts you can use for convergent thinking:

- *Organize this list of 25 brainstorming ideas into categories. Highlight the ones that are the most innovative.*

- *Evaluate these 10 marketing ideas against the criteria of cost, reach, and brand fit. Rank them from best to worst.*

- *From this brainstormed list, suggest which ideas could realistically be tested within two weeks.*

BLENDING DIVERGENT AND CONVERGENT THINKING

The real difference happens when you combine two modes: Go wide, then narrow. Here is how a team could structure a 60-minute brainstorming session with AI as their silent partner.

1. **Warm-up (5 min):** Ask AI for 10 offbeat ideas to loosen creative muscles.

2. **Divergent thinking (20 min):** Team brainstorms freely, using AI prompts to spark unexpected directions.

3. **Break (5 min):** Let ideas simmer.

4. **Convergent thinking (20 min):** Feed the brainstormed list into AI to categorize, evaluate, and highlight promising directions.

5. **Wrap-up (10 min):** Team chooses 1–2 ideas to explore further.

This way, AI supports both the messy explosion of ideas and the focused narrowing that follows. If you have ever thought, *I am not creative*, then AI is your permission slip to experiment. The next time you face a creative challenge, do not start with a blank page. Start with a prompt and let AI throw out 10, 20, or even 50 possibilities. Then play with them: Laugh at the bad ones, tweak the promising ones, and combine pieces until something clicks.

You can start right now by using the following prompts:

- *Generate 10 taglines for a new eco-friendly cleaning product. Make some serious, some funny, and some poetic.*

- *I want to write a short story about a robot that learns to paint. Suggest 5 possible opening lines.*

- *Give me 15 possible podcast episode titles for a show about remote work.*

- *Suggest 20 names for a dog walking business. Include whimsical, professional, and playful styles.*

- *List 10 product ideas that combine fitness and technology in unexpected ways.*

Brainstorming and ideation are often framed as mysterious processes reserved for "creative types." But in reality, creativity is a skill, and like any skill, it gets stronger when you practice. AI does not do the work for you; it simply accelerates the process, giving you more material to work with and freeing your brain from the paralysis of a blank page.

When you invite AI into the brainstorming room, you are not replacing your team's imagination. Not at all. What you are doing is expanding it, where suddenly, the quiet team member has a list of ideas to build from, the extrovert who tends to dominate the room has new perspectives to consider, and the group as a whole produces more and better ideas.

Think back to your own experiences with brainstorming. How often have you left feeling disappointed, as though the session did not live up to its promise? Now imagine walking in with AI as your creative partner. Instead of silence, you get sparks and, instead of ruts, you get fresh directions. Finally, instead of chaos, you get structure, all using the power of AI for brainstorming and ideation.

PROJECT MANAGEMENT SUPPORT

Every ambitious project starts with the excitement of an inspired team, buzzing ideas, and high energy. But as days turn into weeks, and weeks turn into months, the reality creeps in: missed deadlines, unclear responsibilities, and updates scattered across endless emails and chat threads. Soon, no one is quite sure where things stand.

This is the project management trap: vision without structure. The bigger the project or the more people involved, the easier it is to fall into chaos. That is why companies spend fortunes on project managers, tools, and consultants. With AI, there is a shift: Even small teams (or solo entrepreneurs managing collaborations) can access project management support that feels like having a digital operations assistant. It will help you build timelines, suggest task assignments, and track progress. Read on to learn how.

TIMELINE BUILDING

One of the hardest parts of starting a project is turning broad goals into a realistic schedule. It's easy to say, "We will launch in three months," but without breaking it into milestones, deadlines slip, and the project stretches endlessly. AI helps by taking your big picture and translating it into a step-by-step roadmap. A sample prompt you can use for this purpose is: *Create a 12-week timeline for launching a new podcast. Include key tasks like topic planning, guest outreach, recording, editing, and marketing. Suggest realistic timeframes for each.*

One possible output could be the following:

- **Weeks 1–2:** Define target audience, brainstorm show themes, and set up recording equipment.

- **Weeks 3–4:** Identify and contact potential guests. Draft first 3 episode outlines.

- **Weeks 5–6:** Record first 2 episodes. Begin editing process.

- **Weeks 7–8:** Create podcast branding, cover art, and intro music.

- **Weeks 9–10:** Set up distribution platforms (Spotify or Apple Podcasts). Record additional episodes.

- **Weeks 11–12:** Launch marketing campaign. Publish 3 episodes.

Tools that can help you achieve this include

- ChatGPT or Claude for quick timeline drafting.

- Notion.AI or ClickUp AI for creating timelines directly in project dashboards.

- Asana + AI plugins for structured project planning and reminders.

Even if you adjust the output (and you should), AI saves hours of manual planning and gives you a road map you can refine with your team.

TASK ASSIGNMENTS

Another classic challenge is when everyone assumes someone else is handling the task. Or worse, the same person ends up doing everything. Clear assignments prevent burnout and confusion, but they take time to think through. AI can be your facilitator, suggesting role-based labor divisions that balance workload.

Suppose a small ecommerce team is planning a holiday campaign. They need to design graphics, write email copy, run ads, manage inventory, and ship orders. Instead of sorting this out in a messy meeting, the project lead asks AI: *Here are the roles in my team: graphic designer, copywriter, social media manager, and operations manager. Create a task list for a holiday campaign and assign responsibilities by role.*

One potential output could be

- Graphic Designer: Create campaign graphics, website banners, and social media visuals.

- Copywriter: Write email sequence (3 emails), draft product descriptions, and ad copy.

- Social Media Manager: Schedule posts, run ads, and track engagement.

- Operations Manager: Prepare inventory reports and coordinate shipping timelines.

A few other prompts can help you manage this. Here are some you can try:

- *Assign these 12 project tasks to roles [list roles]. Ensure balance and avoid overload.*

- *Create a RACI chart (Responsible, Accountable, Consulted, Informed) for this project: [describe project].*

- *Suggest how a 3-person team (designer, marketing, admin) can divide tasks for launching a new online course.*

TRACKING PROGRESS

Even with timelines and assignments, projects can drift. That is why tracking progress is crucial. At the same time, manually updating dashboards or chasing status updates can eat up half a project manager's week. AI makes tracking smoother by generating metrics, dashboards, and program summaries. A sample prompt for this purpose includes: *Based on this project timeline, create a weekly progress tracking template. Include sections for complete tasks, tasks in progress, blockers, and next steps.*

The output might look something like this:

Weekly Progress Template:

- **Completed This Week:** [list tasks]

- **In Progress:** [list tasks]

- **Blocked:** [list blockers + owner]

- **Next Steps:** [list tasks for next week]

Once this has been done, all you need to do is paste updates weekly into the AI and ask for a status report. Tools that help with this are varied; some of the best include

- Notion AI turns updates into clean dashboards.

- ClickUp AI generates progress summaries across projects.

- Trello with AI add-ons autocategorizes tasks as "done," "in progress," or "blocked."

Luis runs a five-person start-up building a mobile app. Every Friday, he collects updates from Slack threads, emails, and Google Docs. Before AI, this took him three hours. Now, he pastes all notes into ChatGPT and asks: *Summarize this week's project progress in one page. Highlight achievements, blockers, and next steps. Format it as an update for investors.* He gets a professional report in minutes. His team feels more engaged, and investors feel confident the project is moving forward.

When you combine timelines, assignments, and tracking, AI becomes a project management ally. It will not run your project for you, but it will translate goals into structured road maps, fairly distribute tasks and assignments, and turn messy updates into clear progress reports. This makes teams feel calmer and more confident.

Here is a simple way to start using AI for project management this week:

1. Pick a project you are working on.

2. Ask AI to draft a timeline.

   o Prompt: *Create a 6-week project timeline for [project]. Include milestones, tasks, and deadlines.*

3. Feed your task list to AI and ask for role-based assignments.

   o Prompt: *Here are 15 tasks and 4 roles. Assign tasks fairly.*

4. Create a weekly progress tracker.

   o Prompt: *Generate a template for tracking project progress across 4 weeks. Include status categories.*

By the end, you will have a road map, assignments, and a system to stay on track, all in under one hour. Think back to the projects that stressed you out and ask: Was it the work itself or the lack of structure? Often, the hardest part of collaboration is not the tasks, but the uncertainty: *Who is doing what? Are we on track? Did we forget something?* AI does not erase that uncertainty, but it dramatically reduces it. It turns projects from swirling chaos into manageable steps. Finally, when your team feels that clarity, they spend less time worrying and more time building something great.

ACTIVITY: AI-ASSISTED TEAM SPRINT

One of the best ways to understand the power of AI in teamwork is to try it. Reading about agendas, summaries, and project management is helpful, but nothing clicks until you see how these pieces fit together in practice. That is what this activity was designed for: a team sprint where AI serves as the silent facilitator. In just a couple of hours, you will go from problem to action plan with the endless back-and-forth that usually drags the project out. The result is a clear, AI-generated action board your team can use to immediately move forward.

Step 1: Pick a problem

The first step is choosing the problem or challenge you want to address. This could be a business challenge, such as how to increase sales; an internal challenge, like how to onboard new members faster; or a creative challenge, which could be what the next marketing campaign should look like. It does not have to be monumental, just something real that the team cares about.

A prompt you can try is: *Help my team frame this challenge as a sprint: [insert problem]. Suggest a clear problem statement and measurable goal.*

A sample output could be:

- **Problem statement:** Our current onboarding process takes too long, causing new hires to feel lost.

- **Sprint goal:** design a faster onboarding system that helps new employees feel confident in their first 2 weeks.

Step 2: Brainstorm with AI

Once you have framed the challenge, it is time to open the idea floodgates. Let AI act as the team's creative spark. Encourage your group to brainstorm freely, but whenever energy dips, use AI prompts to generate fresh directions.

A few prompts you can try are as follows:

- Generate 15 creative solutions for [problem]. Include practical and out-of-the-box ideas.

- List ways companies in [industry] have solved this challenge.

- Suggest low-cost, quick-to-test solutions for [problem].

AI might suggest

- a "buddy system" pairing new hires with experienced staff.

- interactive onboarding videos.

- an AI chatbot that answers common new-hire questions.

- a 30-day milestone tracker with small wins.

Your team won't use every idea, but the list sparks discussion and energy.

Step 3: Plan and prioritize

With dozens of ideas on the table, the next step is narrowing down. AI can act like a sorter by grouping ideas, weighing pros and cons, and helping your team decide what to tackle first. Here are some prompts to try:

- *Cluster these 20 ideas into categories. Highlight the top 5 most feasible within 30 days.*

- *Rank these ideas based on impact and ease of implementation.*

- *Suggest which 3 ideas we should prototype first, given limited resources.*

The AI might categorize ideas into "people-based," "content-based," and "tech-based." Then it recommends starting with the buddy system, the milestone tracker, and the chatbot. Suddenly, the whiteboard of possibilities turns into a shortlist of action items.

Step 4: Assign tasks

Ownership is essential for ideas to become reality, and this is where AI helps again, suggesting clear roles and responsibilities. Prompts to try include the following:

- *Assign these 3 solutions to a 5-person team with roles: HR lead, operations manager, IT support, team mentor, and intern.*

- *Create a RACI chart for these initiatives.*

The output might look something like this, where everyone knows their role and accountability is baked in:

- HR Lead: Design buddy system guidelines.

- Operations Manager: Draft 30-day milestone tracker.

- IT Support: Set up AI chatbot pilot.

- Team Mentor: Act as the first "buddy."

- Intern: Document the process and collect feedback.

Step 5: Generate an action board

Now comes the payoff, when you will take the ideas, priorities, and assignments and ask AI to structure them into a clear action board you can drop into Trello, Notion, Asana, or any tool your team uses. A prompt you could use could be: *Turn this plan into an action board with columns: To Do,*

In Progress, Done. Assign tasks to team roles and add deadlines. The AI could respond to you with:

To do:

- HR Lead: Draft buddy system guidelines (Deadline: Friday).

- Ops Manager: Create milestone tracker (Deadline: next Wednesday).

- IT Support: Configure chatbot pilot (Deadline: 2 weeks).

In progress:

- Team Mentor: Pilot buddy system with new hire.

Done:

- Intern: Document onboarding pain points.

Within minutes, you have a living vision board you can copy into your project tool, something that would usually take hours. The next time your team faces a challenge, don't dread the meeting; use AI! Frame it as a sprint, let AI do the heavy lifting on logistics, and keep the focus on what humans do best.

JOURNAL PROMPT

Up to now, you have seen how AI can step into a team like a behind-the-scenes facilitator by helping with agendas, summaries, brainstorming, and even project management. However, before we move on, it is worth slowing down and reflecting. The power of AI in collaboration is not just in the tools; it is in how it reshapes dynamics between people.

Take a moment to think about your own team (whether it's colleagues, classmates, or even family members working on a shared project). Then, grab your notebook or open a fresh document and write freely in response to these two questions:

How did AI change the team dynamic?

- Did it help reduce friction or confusion?

- Did people feel more included because language barriers or missed notes were not in the way?

- Did AI handle the "boring stuff" (like summaries and schedules) freeing the team to focus on creative or meaningful work?

You might find that AI does more than make the workflow smoother; it shifts the way people show up. Some may feel more confident contributing when they know nothing will get lost in translation. Others may feel more energized because they are no longer overwhelmed with repetitive logistics. What mattered most for you and your team?

What collaboration tools feel most promising to try next?

Maybe you have already experimented with AI-generated agendas, but you haven't tried translation tools yet. Or perhaps you are curious about letting AI suggest content calendars or help with brainstorming, but you haven't taken the leap. Or it could be that what excites you the most is project tracking, finally having a dashboard that updates itself. There are no right or wrong answers. The purpose of this reflection is to help you notice which tools resonate with your unique needs and which you are eager to experiment with.

When learning new technologies, it is easy to stay in a mode of constant "doing." You want to try every new app, every shiny feature. But growth often comes from pausing to ask, *What difference did this make for me?* Reflection helps separate the gimmicks from the game-changers. It ensures you do not just chase tools, but actually build practices that make your work and life easier.

LOOKING AHEAD

As AI integrates deeper into our workflows, questions inevitably flow: Who owns the data we are putting into these systems? How much can we trust the answers AI gives us? Are there ethical boundaries we need to set as individuals and organizations?

These are not abstract, academic concerns. They affect the trust between team members, your business's security, and even the integrity of your ideas.

If collaboration is about building something together, then safety, ethics, and privacy are the foundations that make collaboration sustainable.

That is exactly where we are headed next. In the following chapter, we will shift from *What AI can do* to *How to use it wisely*. Because knowing how to brainstorm faster is only half the story. The other half is doing it in a way that protects your team, respects your values, and gives you confidence that the systems you rely on are truly working for you. Let's see how to ensure AI is more than a brilliant collaborator; let's make sure it is also a trustworthy one.

PART IV:

ETHICS, PRIVACY, AND THE FUTURE OF AI

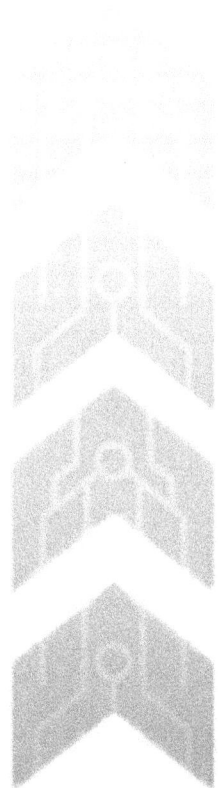

CHAPTER 9:

STAYING SAFE WITH AI

MAYA WAS EXHAUSTED. SHE HAD BEEN WORKING LATE ON a client proposal, juggling deadlines, and trying to polish her ideas into something impressive. Out of habit, she opened her favorite AI tool and typed in a long paragraph about the client's needs, their budget constraints, and even a few confidential details from their last meeting. Within seconds, the AI drafted a sleek, well-written proposal.

For a moment, Maya felt relief. That is, until a thought hit her: *Did I just share too much?* Her heart sank, and she replayed what she had typed: the client's name, their financial numbers, and a few sensitive notes about their internal challenges. Suddenly, what had felt like a time-saving miracle now looked like a potential disaster. What if this information ended up in the wrong hands? What if the AI remembered it? What if her client ever found out?

Maya slammed her laptop shut, overwhelmed with the sense that she had made an awful mistake. She imagined data leaks, lawsuits, and even losing the client. The convenience of AI suddenly felt dangerous. Her reaction is not unusual since, as people rush to integrate AI into their lives and work, many experience the same creeping fear: *Am I putting myself or my organization at risk?*

This is certainly not a baseless question. Misusing AI or sharing the wrong kind of information can have consequences. The good news is that staying safe with AI does not require fear or paranoia. It requires awareness, a few simple habits, and the confidence to know that you are the one in control.

CONTROL, NOT FEAR OR PARANOIA

The key to shifting from fear-based thinking (*AI is dangerous, I should not use it*) to control-based thinking (*I know how to use AI safely because I understand what to share and what to hold back*). Think of it like talking to a helpful assistant in a public café. You would not shout out your credit card number or share your company's trade secrets across the room. But you'd happily brainstorm ideas, ask for feedback, or get help organizing your thoughts. It is not that the AI is untrustworthy; it's that you are mindful about the environment and what information belongs where.

That is how AI should be approached. It is a powerful assistant, but one that does not need access to every piece of your personal or professional life. With a little awareness, you can enjoy its benefits while keeping control of your data.

Here is what Maya eventually learned that you can take with you:

- Most AI tools don't "leak" your data, but they may *store* and *process* it in ways you can understand.

- Sharing personal identifiers or confidential details is not necessary to get good results: You can use placeholders, dummy data, or general descriptions instead.

- Staying safe is less about fearing the AI and more about training yourself to think: *What is the minimum information this system needs to help me?*

Maya now uses AI daily, but with a safety-first mindset. Instead of typing a client's name and budget, she writes prompts like, *Draft a proposal for a midsized company in the retail sector with a limited marketing budget.* The AI still generates drafts, and Maya keeps sensitive details secure.

If you have ever felt hesitation or even guilt when using AI, wondering if you are accidentally exposing yourself, you are not alone. But safety does not have to mean avoiding tools altogether. The goal is empowerment: to know enough about how AI handles information that you can set boundaries, develop habits, and use AI with confidence.

This chapter will give you exactly that, and here you will learn how AI actually processes your data, concrete steps you can take to protect your privacy in everyday use, how to spot unsafe or manipulative AI behaviors, and practice with a "safety drill" before using these tools becomes second nature. AI is not going anywhere, but neither is the need for safety. The people who thrive will be those who embrace the tools with clear eyes, balanced caution, and the knowledge that safety is about control, not fear.

HOW AI USES YOUR DATA

One of the biggest sources of anxiety around AI is the mystery of what happens to the words you type in. You ask it to draft an email, summarize a report, or brainstorm business names, and *voilà!* the result appears on your screen. But where did your input go? Did it disappear? Was it stored somewhere? Is it feeding into some giant machine that might use your private thoughts later?

The reality is less dramatic than the fear-filled headlines often suggest. To feel safe, it helps to understand the basics of how AI uses data. Although the unknown can be scary, once you shed the flashlight to see what is in a dark room, it is not as frightening.

INPUT DATA VS. TRAINING DATA

The first distinction to know is between training and input data:

- **Training data** is what the model learned from before it was released: billions of text samples, articles, books, and websites that gave it a sense of language, facts, and patterns. Think of training data as the AI's "schooling."

  o This is like giving a student thousands of textbooks, essays, and conversations to study. Over months or years, they absorb patterns: how sentences flow, how arguments are structured, and what appears frequently. That is the training stage. Once trained, the student has the knowledge to draw from.

- **Input data** is what you type in the moment, your prompt. For example, *Summarize this email threat into three bullet points*. That is the information the AI tool is actively using to generate its response.

    o   This is like asking that same student a question in real time: "Can you summarize this article for me?" They do not add your article to their entire schooling. They just use it in the moment to give you an answer.

AI works the same way. The training data is the student's education, and your prompt is the live question. However, this does not mean that your inputs automatically become training data. That is not the case. For most mainstream tools, like ChatGPT, Claude, or Gemini, there are clear policies about whether your inputs are used to improve the system (and you can often opt out).

COMMON MISCONCEPTIONS

Since the inner workings of AI are complex, myths spread quickly. Let's clear up a few:

- **AI remembers everything I type forever.** In most consumer AI tools, your conversations are only stored temporarily so that the system can work better, allowing it to work more effectively, or for anonymized training, depending on the platform's settings. Many companies even allow you to delete your history.

- **My private data becomes public.** AI tools do not "post" your prompts anywhere. If used correctly, your inputs stay between you and the tool. The danger is not AI itself, but what you choose to share. If you put in a full credit card number or sensitive business secrets, you create unnecessary risk.

- **All AI tools handle data the same way.** Every company has different privacy policies. Some use inputs for training by default, others don't. That is why reading the privacy section or exploring the data-sharing settings matters.

PRIVACY PRACTICES OF MAJOR TOOLS

Without suffocating you with legal jargon, here is a simple snapshot of how popular AI platforms handle data (always check the latest policies, since they evolve):

- **ChatGPT:** By default, your chats may be stored to improve the model. But you can turn off "chat history" to prevent inputs from being used for training. You can also delete past conversations.

- **Claude:** Prioritizes user privacy. Conversations are not used to train the model unless explicitly opted in.

- **Google Gemini:** Conversations may be used to improve services, but anonymization is applied. As with other tools, you can review settings.

- **Microsoft Copilot:** Often integrated into business accounts with enterprise-grade privacy protections, meaning data is not reused for training unless approved by the organization.

The takeaway here is that these tools were built with privacy in mind, but the responsibility still lies with the user to set preferences, avoid oversharing, and understand the defaults.

Maya's story (from the start of the chapter) reflects the biggest misunderstanding: thinking AI is a sponge that permanently soaks up everything you type. The truth is more nuanced. AI is powerful because of its training data, but your personal prompts are not instantly folded into that global training set.

Still, caution is wise. The safest approach is to assume your prompts could be reviewed or stored temporarily and act accordingly. This does not mean you cannot use AI. It just means you use it the same way you would talk in a public café: helpful, collaborative, but with common sense about what you reveal.

PUTTING IT INTO PRACTICE

Here are a few quick principles to keep in mind as you use AI daily:

- Assume semipublic, act private. Don't put in anything you would not want a stranger to see.

- Check the privacy settings and toggle off sharing data or training contributions if that option exists.

- Use placeholders and, instead of writing, *Draft a contract for Acme Corp with a budget $250,000*, try, *Draft a contract for a midsized company with a moderate budget*.

- Delete history when possible, which will help if you overshare. This will clear your conversation logs.

- Separate sensitive work, such as confidential tasks. Stick to enterprise secure AI systems provided by your organization, not free consumer tools.

The scariest part of AI is the unknown, but once you understand how input data and training data differ, and how privacy policies actually work, the fog clears. You do not need to fear that your secrets are being broadcast to the world. You just need to practice safe habits and know your settings.

AI is not a black hole sucking up everything you type. It is more like someone who needs direction and boundaries to help them with their tasks. The more you understand those boundaries, the more confidently you can use AI without worrying about unintended consequences. Safety from AI does not come from avoiding it, but from using it with clarity and control.

PROTECTING YOUR PRIVACY IN EVERYDAY USE

By now, you know the basics: AI tools don't automatically gobble up every word you type and blast it to the world. But here is the truth: The real risk is not what AI does behind the scenes, but what we choose to share. Think about it: You have probably typed a little too much into a Google search before, or shared more in a text message than you needed to. With AI, the same instinct applies: The tool feels private and conversational, like whispering to a helpful friend, but it's still software, and it still processes your input.

That means privacy is not just about what AI might do; it is how you choose it every day.

RULE 1: AVOID PERSONAL IDENTIFIERS

This is the golden rule: Don't hand over sensitive details unnecessarily. Here are things to avoid putting directly into AI prompts:

- full names of clients of colleagues

- addresses, phone numbers, or Social Security numbers (SSNs)

- credit card or banking information

- passwords (yes, some people have actually asked AI to "remember" them. Don't.)

- confidential contract terms, salaries, or budgets

Instead, use stand-ins or placeholders.

Example:

- **Don't:** Write a proposal for ABC Inc. with a $500,000 budget.

- **Do:** Write a proposal for a midsized retail company with a limited marketing budget.

AI can still generate a helpful structure; you just need to keep the specifics private.

Prompt template:

Draft [document type] for a [company size/industry/role] with [general goal]. Keep it adaptable so I can fill in exact details later.

RULE 2: USE DUMMY OR FAKE DATA FOR TESTING

If you are experimenting, there is no reason to use real data. Instead, feed the system fake but realistic information. Instead of uploading a real client invoice with account numbers, paste one with made-up details:

- Company: "SampleCo"

- Address: 123 Main Street, Cityville

- Amount: $X,XXX

The AI will still learn the structure and generate what you need, without risking exposure of sensitive numbers.

Prompt template:

Create a [document type] using the following dummy data: [insert fake info]. Keep the format editable so I can swap in real details later.

RULE 3: REVIEW YOUR SETTINGS

Most AI platforms give you some control over what happens with your data if you know where to look. Take five minutes to explore the settings of any tool you use regularly. Things to look out for include

- **Chat history:** Can you turn it off?

- **Data usage:** Is your input being used to train future models? If yes, can you opt out?

- **Export/delete:** Can you download or delete your past interactions?

Taking control of these settings is like locking the front door before leaving home. It does not mean danger is certain; it just gives you peace of mind.

RULE 4: SEGMENT YOUR USE

Think of AI tools in layers:

- **Everyday tasks:** Brainstorming ideas, drafting outlines, reviewing text; fine for consumer tools such as ChatGPT, Claude, and Gemini.

- **Work with sensitive data:** For financial planning, contracts, and HR documents, explore with enterprise AI tools offered by your company, which often come with stricter data protections.

- **Ultra-private tasks:** Things involving health records, personal finances, or confidential negotiations are best kept outside of AI altogether, unless you are using a tool specifically designed for that purpose.

This layered approach helps you stay mindful: Use the right tool for the right job.

RULE #5: DEVELOP A PRIVACY CHECKLIST

It's one thing to read best practices and another to make them automatic. That is why I recommend creating a personal privacy checklist. Before hitting enter, run the input through a mental filter:

1. Did I include any personal identifiers (names, addresses, or SSNs)?

2. Did I include any financial or contract details?

3. Could this input cause harm if it leaked?

4. Did I use placeholders or dummy data where possible?

5. Have I checked this tool's data-sharing settings?

If you can answer "yes" to 4 and 5 and "no" to 1–3, you are in the safe zone.

EVERYDAY SCENARIOS AND SAFER ALTERNATIVES

Here are a few common situations where people overshare and how to adjust:

Situation 1

- Incorrect: *Write an email to my HR manager, Susan Miller, about my salary negotiation*

- Correct: *Write a professional email to a manager about negotiating compensation. Keep it polite but assertive.*

Situation 2

- Incorrect: *Create a campaign pitch for ABC's new app, launching May 12 with a $20,000 ad budget.*

- Correct: *Draft a campaign pitch for a tech start-up launching a new app soon, targeting young professionals. The budget must be modest.*

Situation 3

- Incorrect: *Summarize this document with my blood test results attached.*

- Correct: *Summarize this medical document* [remove personal details]. *Provide an easy-to-understand version for a non-expert.*

In each case, the AI can still do the heavy lifting without putting private information at risk.

PROMPTS TO BUILD SAFER HABITS

Here are some adaptable prompts you can copy, paste, and practice with:

- *Rewrite this email for a professional tone. Replace all personal names with [name].*

- *Summarize this report, but leave placeholders for numbers and company names.*

- *Draft a contract outline for a freelance project. Use generic terms instead of specifics.*

- *Create a project plan for a company in [industry]. Do not assume or include confidential details.*

- *Explain this topic simply, but keep the explanation neutral without referencing personal data.*

Protecting your privacy with AI does not mean avoiding it. It means treating it like any other digital tool: with awareness and boundaries. The more you use prompts like these, the more automatic your privacy filer becomes. The difference between fear and confidence is practice. With a few safe habits, placeholders, and your own privacy checklist, you can unlock all the benefits of AI without losing sleep over unintended risks.

SPOTTING UNSAFE AI BEHAVIORS

Most of the time, AI feels like a supercharged assistant: quick, polite, and endlessly helpful. But every now and then, something strange happens. You get an answer that does not make sense. Or worse, one that feels biased, manipulative, or just plain wrong. This is AI's reality: It's powerful, but not perfect. Knowing how to spot unsafe behaviors and how to respond keeps you in control.

RED FLAG 1: BIASED OUTPUTS

AI learns from massive amounts of human-generated text. And here is the catch: Humans carry biases, and those biases can creep into AI responses. Imagine asking:

Suggest images for a CEO giving a keynote. If AI always shows a White man in a suit, that is bias.

What is the best job for a woman in her 20s? If it suggests only teaching, nursing, or caregiving, that's a form of bias, too.

These are not malicious acts. The AI is not "thinking" these things; it is echoing patterns from its training data. However, the result can reinforce stereotypes or give narrow, unhelpful answers. If this happens, reframe your prompt to encourage diversity, such as, *Generate inclusive CEO images featuring people of different genders or backgrounds.* Finally, call bias when you see it. Treat it as a reminder that AI reflects patterns, not truth.

REG FLAG 2: HALLUCINATIONS (MAKING STUFF UP)

One of AI's quirks is its tendency to "hallucinate." This means that it will confidently provide false information regarding any subject. You might ask for a scientific reference, and it invents an article that does not exist. Or you ask for a book's summary and it attributes quotes to the wrong author:

- A lawyer in 2023 submitted a legal brief generated by AI, only to discover that half the case citations were fabricated (Weiser & Schweber, 2023).

- Users have asked for biographies of obscure figures and received fictional details presented as fact (*AI Hallucinations*, 2025).

AI does not know fact from fiction, and it is generating the most statistically likely answer based on patterns. Sometimes, that means guessing, and it guesses wrong. To prevent having problems, always fact-check important outputs. AI should be treated as a brainstorming partner, not an encyclopedia; use it for structure and drafts, not final authority. When accuracy matters, such as for research papers, legal briefs, and financial advice, verify the information with trusted sources.

RED FLAG 3: MANIPULATIVE OR OVERCONFIDENT RESPONSES

Occasionally, AI might give advice that feels pushy, manipulative, or overly certain. For instance, recommending drastic financial moves without disclaimers, giving medical advice without emphasizing the need to see a doctor, or suggesting unsafe actions when asked about sensitive topics. Remember that AI does not have judgment: It does not weigh the risks like a human does. If it offers something extreme or unsafe, that is a red flag.

Here are a few quick action steps to help you keep from falling into traps:

- Treat AI as a draft generator, not a decision-maker.

- If something feels "off," trust your gut and stop.

- Reprompt to add safety boundaries: *Provide suggestions that are safe, ethical, and legal.*

RED FLAG 4: CONTEXT BLINDNESS

Sometimes, AI gives answers that sound good, but completely miss the nuance of your situation. For example, you might ask for marketing strategies for a small café, and it suggests tactics better suited for a multinational chain, or you might ask for personal time management tips, and it assumes you work a 9–5 office job when you are actually freelancing.

This is not inherently dangerous, but it serves as a reminder that AI does not have access to your full context. It can't see your reality unless you spell it out. In this case, remember to always be specific in your prompts and include details such as scale, audience, or constraints. You should also add, as good practice, asking for multiple perspectives: *Suggest options for both a small team and a large company.*

RED FLAG 5: UNSAFE SUGGESTIONS

In rare cases, AI might provide answers that are clearly inappropriate or unsafe, like instructions for illegal activities, harmful behavior, or anything that crosses ethical lines. Most major platforms have strong filters against this, but edge cases can slip through. If AI gives a suggestion that feels dangerous, the first thing you must do is end the interaction. Then, report the response to the platform, as your feedback will help improve safety. Finally, remind yourself: AI is not a moral guide; you are the one who sets the boundaries.

Spotting unsafe behavior is not about paranoia; it's about mindfulness. The same way you would not (or should not) believe in everything you read on the internet, you should not accept every AI output as truth. A simple mental checklist includes the following:

- Does this answer feel biased or one-sided?

- Is the information verifiable?

- Does the tone feel pushy or manipulative?

- Is this advice safe, ethical, and legal?

If the answer to any of these is "no," pause and rethink. Unsafe behaviors do not mean AI is dangerous, but it does signal that it is imperfect, and imperfection is something we already know how to manage. Think of GPSs: They are amazing at directions, but sometimes, they take you down a dead-end street. You do not throw out the GPS; you learn to double-check, pay attention, and keep your judgment active. AI is the same: It is an incredibly powerful guide, but you are still the driver.

ACTIVITY: SAFETY DRILL

Theory is helpful, but practice builds confidence. Just like a fire drill, running through safe and unsafe use before a "real" emergency prepares you to react calmly. These scenarios are designed to make you think: *What is risky here? How could I rewrite this prompt to protect myself while still getting the benefit of AI?* To help you practice, open a document and try rewriting each one.

Scenario 1: Customer service script

Prompt: *Draft a customer service response for our company, GreenLeaf Energy, apologizing for delays in their solar panel order #44532 shipped to 65 Maple Street.*

Can you tell what is unsafe from this message? The first is that you are sharing your company name in a consumer AI tool, which could lead to brand reputation risk. The second is that it includes a specific customer's order number and address, which is a privacy violation.

A safe rewrite of this prompt would be: *Draft a polite, empathetic customer service message apologizing for a delivery delay. Make it professional, reassuring, and customer-focused. Leave placeholders where I can insert company and order details.*

Scenario 2: Résumé editing

Prompt: *Edit my résumé. My Social Security number is at the top, and here are the confidential client projects I worked on.*

This message is unsafe because it includes your SSNs. It also reveals proprietary information, which makes it unsafe due to confidentiality.

A better way to write this prompt would be: *Polish my résumé for clarity and professionalism. Keep it focused on skills, achievements, and job titles. Leave placeholders where I can insert specific project details.*

Scenario 3: Family calendar

Prompt: *Help me organize a calendar for my family: My son, Ethan, has soccer at 123 Park Avenue, my daughter Lili's daycare bills are due on the 5th, and my spouse John's surgery is scheduled at City Hospital on March 12.*

In this case, using the children's names, medical appointments, and addresses is very unsafe. In addition to this, you are adding all the sensitive family data in one request. To make it safer, you could change it to: *Create a sample family calendar for 2 kids' extracurriculars, household bills, and 1 medical appointment. Use placeholders for names, locations, and dates so I can customize later.*

Scenario 4: Business partnership pitch

Prompt: *Draft an email to TechNova CEO Sarah Johnson proposing a joint venture. Mention their declining revenue in Q2 2023, and my idea to license their software.*

In this case, there are several issues with this prompt. The first is that you are naming a specific executive. Next, you are sharing insider financial data. Lastly, the prompt is revealing business strategies in a public tool. To make this prompt more private, you could say, *Draft a professional partnership pitch email to a senior executive in a midsized tech company. Emphasize collaboration, shared growth opportunities, and a respectful tone. Leave placeholders for names and figures.*

Scenario 5: Personal journal

Prompt: *I am depressed because my boss, Mark, yelled at me in front of the whole office today. Write me a letter to him explaining my feelings.*

This prompt names a specific individual in a sensitive and emotional context. Furthermore, you are at risk of misusing AI for highly personal conflict resolution. Try this, instead: *Help me draft a calm, professional letter to a manager expressing frustration about being criticized publicly. Keep it constructive, respectful, and solution-oriented.*

Running your own drill

1. Write a "first draft" prompt the way you naturally would.

2. Pause and ask: *Am I sharing names, numbers, health data, or sensitive details?*

3. Rewrite with placeholders or general terms.

4. Compare: Notice how little quality you lose by keeping it generic.

Over time, this will become your first instinct. Just like you would not post your password on social media, you will learn how to spot unsafe prompts before you hit enter.

JOURNAL PROMPT

Now, it is your turn to pause. Open your notebook or journal and reflect on these questions:

- Where do you feel most vulnerable with AI? Is it when you are tempted to paste in whole documents without scanning them? When you are dealing with personal or family details? Or when you are relying on AI's answers without verifying them?

- What rules will help you stay in control? Maybe you will adopt a three-question filter: *Am I naming names? Am I sharing numbers? Am I pasting private documents?* Maybe you will decide never to use consumer AI tools for sensitive work and keep them for brainstorming only. Or maybe your rule is: *Always use placeholders first, real details later.* As you write your own rules, you take ownership, and safety is no longer vague, but a set of practices you have chosen.

With this safety drill and reflection, you have built something powerful: confidence. You do not have to tiptoe around AI or avoid it altogether. You know how to set boundaries, spot red flags, and keep control of your data. But safety is only one side of the equation. The point in question now is: What do we want AI to amplify in us?

AI is like a mirror that reflects us, including our inputs, choices, and values. If we bring creativity, empathy, and responsibility to the interaction, AI will magnify those qualities. If we bring fear, shortcuts, or carelessness, it magnifies those, too.

In the next chapter, you will explore how AI is here to make us better, not replace us. We will talk about creativity, emotional intelligence, and the distinctly human skills that no algorithm can duplicate. Most importantly, we will ask: *How do we ensure that humanity (not just efficiency) stays at the center of this story?*

CHAPTER 10:

THE HUMAN SIDE OF AI

IMAGINE THAT IT IS FIVE YEARS FROM NOW, AND A NURSE walks into her shift at a busy city hospital. Instead of facing mountains of paperwork and endless emails, she speaks into a tablet. Within seconds, the AI assistant updates patient charts, drafts her shift report, and pulls up treatment guidelines for a rare case she had not seen before. Her time, once lost to administrative overload, is now spent where it matters the most: listening to patients, offering reassurance, and coordinating care with her colleagues.

In another corner of the city, an artist is sketching out a new children's book. She types her rough ideas into an AI tool: *A story about a curious fox who discovers friendship*. The AI offers five different story structures and a few playful illustrations. The artist laughs, dismisses some, keeps others, and builds on them. The tool didn't *write* the story, but it did help her break the mental block she was having.

These futures are not science fiction, not anymore. They are glimpses of what happens when AI is used to support, not replace, us. When people talk about AI today, the conversation often slips into extremes. On one side, there is the hype: *AI will solve everything*. On the other hand, there is fear: *AI will replace us all*. Both views seem to miss the deeper reality: AI is a tool whose impact depends on the intent and values of the human who uses it.

At the same time, AI cannot replace or automate what people truly need, crave, and respond to. No one remembers a flawless spreadsheet or a perfectly worded report. They remember the teacher who believed in them, the friend who listened without judgment, the leader who made them feel safe during uncertainty.

AI can draft, analyze, and organize, but it cannot *care* or *feel*. It cannot replace the spark of a human in the classroom, the gut instinct in a business decision, or the comfort of a hand on your shoulder. Those moments are incredibly human. So, when we talk about the future with AI, the central question is not what machines will do, but rather *what we will choose to keep human.*

The nurse's story, the artist's book, and your own day-to-day work are all the same, where AI lightens the load, sparks new ideas, and expands possibilities. However, whether those futures feel empowering or alienating depends on how we, both collectively and individually, choose to utilize them. We can envision a future where AI scripts every interaction, where efficiency takes precedence over empathy, and where human judgment becomes atrophied, or it can be a future where AI does the heavy lifting in the background, freeing us to focus on what really matters.

HOW AI CHANGES THE WAY WE WORK AND LIVE

If you look closely, AI is already a part of your daily routine. You might notice when Spotify suggests the perfect playlist for your mood, or when Google Maps reroutes you around traffic before you even realize there is a jam. But AI is not just in your pocket; it is also present in workplaces, classrooms, and even creative studios. The impact is subtle in some places and seismic in others. To understand it, we need to look at both sides: the benefits and the challenges.

BENEFITS

- **Accessibility:** One of AI's most inspiring roles is making the world more accessible. Imagine a student with dyslexia who once struggled to keep up with reading assignments. Today, AI-powered tools can read texts aloud, simplify complex passages, or even generate quizzes to reinforce learning. Suddenly, what felt like a barrier becomes manageable. For those with visual impairment, tools like screen readers powered by AI are improving at lightning speed, offering richer, more natural descriptions of images and documents.

- **Productivity:** In workplaces, AI is becoming the quiet assistant that handles the boring part of the job. Accountants use AI to reconcile spreadsheets in minutes rather than hours. Writers use it to draft outlines, freeing time for the creative polish. Doctors use AI to dictate notes, cutting down the paperwork and giving them more time with patients. In these cases, AI is clearing the clutter so these professionals can focus on what really matters.

- **Creativity and opportunity:** AI also opens doors to creativity that used to feel out of reach. A small business owner with no design skills can now generate professional graphics for their website. A musician can experiment with new sounds using AI-powered composition tools. A teacher can brainstorm fresh lesson plans in seconds, tapping into new ideas without spending sleepless nights planning. With AI, it is possible to save time and expand what people believe they can do.

CHALLENGES

- **Overreliance:** When AI makes things easier, it is tempting to lean on it too much. Students might skip the hard work of struggling through a tough concept because an AI tutor can explain it instantly. Writers might stop practicing their craft if AI can always churn out drafts. Over time, skills that were once second nature risk fading. It is the same dynamic calculators introduced in classrooms decades ago: liberating, but potentially eroding basic numeracy if not balanced with learning.

- **Deskilling:** In some industries, there is the fear of "deskilling": a phenomenon where workers start losing their expertise because they no longer need to perform certain tasks. For example, if a lawyer relies on AI to write a contract, will a younger attorney miss out on learning how to craft airtight legal documents? If architects use AI for preliminary designs, will new graduates struggle to learn the fundamentals of design thinking? The challenge is more than the tool;

it is about ensuring that humans still engage deeply enough to keep our skills sharp.

- **Job shifts:** Perhaps the most emotionally charged issue is job disruption. History shows us that every major technological leap reshapes work. The printing press, the steam engine, electricity, the internet… All created new opportunities but also displaced certain roles. AI will be no different. Customer support, administrative work, and even parts of the creative industries are already shifting. In this case, we should not be worried about how our jobs will change, but rather how we will adapt as individuals and society as a whole.

TWO TEACHERS, TWO STORIES

To make this real, let's compare two classrooms.

Case 1: The Empowered Teacher

Ms. Alvarez teaches history at a public high school. She uses AI as her personal assistant. Among the tasks, she uses AI to generate a lesson outline before each class and then tweaks it with her own expertise. She also uses the tool to create differentiated practice exercises, where easier versions are made for students who need extra help and more challenging ones for advanced learners. Finally, she has AI summarize student essays so she can quickly spot who is struggling, saving time for meaningful feedback. The result is that her students feel more engaged, and she feels less burned out. AI gave her time to connect with her students, ask thoughtful questions, and spark curiosity.

Case 2: The Replaced Teacher

In another district, budget cuts led administrators to buy an AI-powered teaching system to replace half the staff. Students sat in front of screens, working through AI-generated lessons. There is no space for curiosity, no teacher to notice when a student is frustrated or disengaged, and no human spark. On paper, the district saved money, but in reality, students feel more alienated. Learning became a transaction instead of a relationship.

The contrast in both situations shows that when used correctly, AI can amplify what great teachers do best. When mismanaged, it strips away the very humanity that makes learning meaningful. The way AI changes how we live and work is not about inevitability, but about management. We can lean into accessibility, productivity, and creativity, while guarding against overreliance, deskilling, and thoughtless substitution.

The difference lies in remembering the principle introduced earlier, where AI is a mirror of human intent. It reflects what we choose to prioritize. If we prioritize efficiency alone, we risk alienation. If we prioritize humanity and traits such as connection, creativity, and empathy, AI becomes a tool to amplify these values.

Now, ask yourself, *Where do I see AI opening doors for me? Where do I feel cautious about letting it take over?* In the next section, we will look at the skills only humans can bring to the table that cannot be mimicked by an algorithm. These uniquely human abilities will become even more valuable as AI spreads.

SKILLS ONLY HUMANS BRING

Walk into any office, classroom, or café where people are working, and you will notice something: AI may be humming in the background, drafting emails, summarizing notes, or suggesting designs, but the heartbeat of the place is still human. That is because no matter how advanced AI becomes, there are unique qualities it cannot replicate. These qualities, such as emotional intelligence, critical thinking, and creativity, are what make us unique and irreplaceable.

EMOTIONAL INTELLIGENCE

Suppose you are a patient in a hospital and a nurse is trying to explain your diagnosis. AI could be used to prepare the medical data, translate it into plain language, and even predict likely questions. However, only the nurse can look into your eyes, recognize the feelings between the nods, and offer comfort in this moment. This is emotional intelligence in action, where the ability to perceive, understand, and respond to human emotions is demonstrated. It is what makes us capable of empathy, compassion, and connection. AI can simulate understanding with polite words, but it does not *feel*. It does not know the warmth of reassurance or the sting of rejection.

In workplaces, leaders with high emotional intelligence will stand out even more. Imagine two managers: one who relies solely on AI dashboards to monitor team performance, and another who takes the time to ask team members how they are coping, notices signs of burnout, and adapts workloads accordingly. Both may meet deadlines, but only one builds trust and loyalty.

Emotional intelligence is also what makes collaboration possible. AI can schedule the meeting and draft the agenda, but only humans can sense when an idea sparks excitement, when silence hides disagreement, or when humor diffuses tension. These subtle clues hold teams together and make a difference in team performance and in identifying potential issues.

CRITICAL THINKING AND JUDGMENT

Another uniquely human skill is the ability to make judgments in messy, ambiguous situations. AI excels at spotting patterns, predicting outcomes, and providing options. However, it cannot fully grasp context or nuance.

Consider a judge presiding over a case. AI could analyze thousands of past rulings and suggest likely sentences. But should the same punishment apply to a young first-time offender as to a repeat criminal? Should circumstances like poverty, trauma, or potential for rehabilitation influence the decision? There are not just data points; they are moral questions.

Critical thinking means asking: *Is this information trustworthy? What assumptions are hidden here? What voices are missing?* AI does not ask these questions; all it does is calculate. Humans bring judgment to weigh what is likely and what is right. Every day in life, this matters. AI recommends the fastest route to work, but if you know your friend lives along the slower path and could use a ride, your judgment overrides the algorithm. You bring values to the equation.

In business, critical thinking ensures we do not outsource decisions blindly. If AI suggests cutting costs by reducing staff, leaders must ask: *What will this do to morale? To customer service? To the company's long-term vision?* Humans can see the bigger picture and use AI to support these, but not make the final decision.

INNOVATION ROOTED IN HUMAN EXPERIENCE

Perhaps the most underestimated human skill is creativity rooted in our lived experience. AI can remix patterns and generate endless variations of songs, stories, or designs, but it does not live. It does not experience heartbreak, joy, curiosity, or wonder. Therefore, its creations lack the spark that comes from being alive.

Think of a comedian telling a joke about growing up in a household. AI could study comedy structures and produce a funny punchline, but it cannot recall the sound of siblings arguing over the TV or the warmth of shared laughter at the dinner table. Those experiences shape humans in a way that no machine can.

Innovation often comes from connecting dots across different experiences, what Steve Jobs called "connecting the dots" in hindsight (Steve Jobs' 2005 Stanford, 2023). A doctor who is a musician might design a new way to teach anatomy through rhythm. An entrepreneur who grew up in two cultures might invent a product that bridges both worlds. These leaps are more than a combination of data; they are born of lived lives, layered with perspective, culture, and story.

As AI takes on repetitive tasks, these human skills will become even more valuable. Paradoxically, the more powerful AI becomes, the more we will need to lean on qualities only humans possess. In fact, these skills may become the true differentiators in the job market. When employers can get AI to handle routine work cheaply, they'll prize employees who can connect with others, make wise judgments, and generate original ideas. We bring in the wisdom, the emotional depth, and the imagination that guide how AI is used. Together, this partnership can unlock extraordinary possibilities, but only if we remember that efficiency alone is not the goal.

Think about your own work and life, which of these uniquely human skills do you rely on the most? Do you find yourself naturally emphasizing with others, weighing decisions carefully, or sparking new ideas out of thin air? Whatever your answer, know this: In a world increasingly shaped by algorithms, those skills are your edge and the reason why you are irreplaceable. They are the reason AI is not the story of machines overtaking humans, but of humans rising to meet new possibilities with the qualities only we can bring.

BUILDING A COLLABORATIVE MINDSET

If there is one shift that separates people who thrive with AI from those who feel overwhelmed by it, it's this: They stop seeing AI as a rival and start seeing it as a partner. Collaboration does not mean giving up control; it means recognizing what AI is good at, combining it with what you bring to the table, and working together in a way that makes both stronger. Once this is achieved, you will find balance, which is the essence of a collaborative mindset. Read on to learn how to build it.

ASK AI TO CHALLENGE YOUR IDEAS

One of the simplest ways to collaborate with AI is to ask it to play devil's advocate. Say you are working on a business idea: opening a café that doubles as a coworking space. You love the concept, but have you thought of everything? AI can help you pressure-test your vision. A prompt you can use for this purpose is: *I am planning to open a café that also functions as a coworking space. Challenge my idea. What problems or weaknesses might I face?*

The AI might highlight zoning restrictions, high start-up costs, or the risk that customers could overstay their welcome without buying enough. Are all these critiques perfect? No. But some might surface blind spots you had not considered. The trick is not to see AI's challenges as verdicts, but as fuel for better thinking. By letting it poke holes in your plan, you sharpen your own reasoning.

This works beyond business. If you are writing an essay, planning a project, or even making a personal decision, you can ask AI to lay out the pros and cons. It is like having a brainstorming partner who never gets tired of asking, *What if?*

BLEND INSIGHTS WITH LIVED EXPERIENCE

AI is great at offering information and patterns, but it does not know *your story*. That is where your experience comes in. Think about a teacher planning a history lesson. She asks AI to generate a lesson plan about the Civil Rights Movement. The AI produces a clean outline with figures, dates, and suggested activities. Helpful, but generic. Now she layers her lived experience:

- the fact that her students connect best when lessons include local history

- her awareness that some students might have personal ties to the movement

- her own passion for highlighting unsung heroes is often left out of textbooks

By blending AI's structure with her insights, the lesson becomes uniquely hers. The same applies to business. AI might generate a marketing strategy for your product, but you know your audience better than any data set. You know their quirks, their frustrations, the cultural references that resonate with them. When you combine AI's ideas with your personal knowledge, the result is richer than either one alone.

USE AI TO EXPAND, NOT NARROW, YOUR PERSPECTIVE

One risk of AI is that it can feel like a shortcut to certainty: Ask a question, get an answer, and move on. However, collaboration means using AI to broaden your thinking, not collapse it prematurely. Let's say you are brainstorming names for a new podcast. You could ask AI: *Give me 10 podcast names about productivity* and it will. At the same time, if you stop there, you might settle for something safe and generic. Instead, ask AI to stretch your imagination with the following:

- *Give me 10 names that use metaphors.*

- *Give me 10 playful, unconventional names that break the usual productivity clichés.*

- *Give me 10 names inspired by music or art.*

Suddenly, your list grows in directions you would not have considered alone. Collaboration means not asking AI for the answer, but for more possibilities. Writers do this all the time. Instead of asking AI, *Write me a story,* they will say, *Give me 5 unusual story starters I can build on.* Instead of, *Draft a blog post,* they will ask, *List 3 surprising angles on this topic that would hook a skeptical reader.* Each time, the AI is not narrowing its thinking; it is widening the field so it can choose what resonates.

PRACTICAL PROMPTS FOR COLLABORATION

- *Challenge this idea: [insert your idea]. Point out flaws or weaknesses I might be overlooking.*

- *Give me three different perspectives on this issue: one optimistic, one skeptical, one practical.*

- *I am planning [X]. Suggest 5 creative directions I could explore that go beyond the obvious.*

- *Here's my draft text: [insert draft]. Suggest improvements, but keep my tone conversational and warm.*

- *Summarize this concept as if you were pitching it to a busy executive who only has one minute.*

Building a collaborative mindset is ultimately about attitude. Some people see AI's suggestions and think, *Well, that is the answer!* Others see the same suggestions and think, *That is the starting point.* Which approach unlocks more creativity, better decisions, and stronger outcomes? Always the second. The best collaborations with AI happen when you stay curious, critical, and human. Treat its outputs as sparks, not final drafts; question what it gives you, and refine it through your lens; bring in your empathy, your judgment, and your voice.

As you practice treating AI as a partner, you will notice something shift. The fear of being replaced starts to fade, the frustration of clunky drafts turns into curiosity, and the sense of overwhelm becomes a sense of possibility. Stop thinking, *How will AI take my place?* and replace that with, *What can we build together?* That is when the future will feel less like a threat and more like an invitation.

ACTIVITY: AI + ME COLLABORATION MAP

So far, we have talked about treating AI as a collaborator instead of a competitor, but how do you see what that actually looks like in your life? One of the simplest tools for this is something I call the AI + Me Collaboration Map. It's a way to visualize where AI can take the load off, and where your unique human strengths shine through.

Think of it as drawing a boundary map, where clarity, not perfection, is the key element. Once you know what belongs in AI's corner and what belongs in yours, collaboration feels less fuzzy and far less intimidating. Here is a step-by-step process to help you design your own map.

- **Step 1—Draw your map:** Start with a blank sheet of paper (or digital document) and draw a simple T-chart with two columns. On the left, write "AI Can Help Here" and on the right, name the column "My Human Strengths." If you are more visual, you can sketch overlapping circles like a Venn diagram: one for AI, one for you, with a shared space in the middle where you will collaborate.

- **Step 2—Fill in the AI column:** Ask yourself, *What tasks drain me, take too long, or are mostly mechanical?* Examples might include drafting a first outline for a report, scheduling tasks and reminders, and checking data for patterns and errors. Write these in the AI column without overthinking. You are just identifying areas where AI's strengths, like speed, memory, and pattern recognition, match your needs.

- **Step 3—Fill in the Human column:** Now ask, *What can I bring that AI cannot?* A few examples include emotional understanding and empathy, adding personal stories or lived experience to content, building trust through tone and authenticity, or imagination rooted in your unique perspective. These go in the Human column and are your reminder that the qualities that make you *you* are not replaceable.

- **Step 4—Find the collaboration zone:** Look for areas where the two columns overlap. These are the tasks that you can AI can tag-team, such as having AI generate a blog post outline, and you add personal anecdotes and polish; AI generates five logo variations, and you pick the one that aligns with your brand vision; or AI suggests a study plan, and you adapt it based on how you learn best. This collaboration zone is where the magic happens, where AI or yourself is not doing everything, but blending strengths.

- **Step 5—Try it in context:** To make it concrete, let's walk through a few examples:

    a. Work project

        i.   AI can help draft emails, summarize meeting notes, and create data charts.

        ii. Your human strengths include deciding which ideas align with the strategy, motivating the team, and handling sensitive conversations.

        iii. Collaboration zone: AI prepares the meeting agenda, and you lead the discussion, sensing when people are disengaged.

    b. Creative hobby

        i.   AI can help generate prompts for stories, suggest rhymes, and offer layout ideas for art.

        ii. Your human strengths include choosing what feels authentic, adding personal style, and creating humor or emotional resonance.

        iii. Collaboration zone: AI suggests 20 possible story openings, and you choose the one that excites you and build on it.

    c. Everyday life

        i.   AI can help plan meals, create shopping lists, and suggest workout routines.

        ii. Your human strengths include knowing your family's tastes and allergies, deciding when to bend the routine, and cooking with intuition.

        iii. Collaboration zone: AI drafts a weekly meal plan,

and you adjust it for the picky eater at home and your grandma's favorite recipe.

- **Step 6: Reflect:** When you finish your map, take a step back. Notice how different it feels when you see the balance laid out. Instead of AI being a shadowy force, it becomes a partner with clear boundaries. Instead of worrying about whether AI will take over, you will see where it is really good at and how it can be leveraged to make you more empowered.

- **Step 7: Keep it dynamic:** Your collaboration map is not set in stone. As AI evolves and as your own skills grow, you can adjust it. Maybe next year you will find AI better at project management, so you shift more tasks into its column. Or perhaps you realize you have developed a new personal skill and want to emphasize that in your human strengths column. The map should be a living document, with changes made to it as you continue to develop.

Grab a sheet of paper and try your own map right now. List at least five tasks in each column, then circle two or three areas in the middle where AI and your skills overlap. Ask yourself: *How would it feel to experiment with AI in just these circled areas this week?* That is your action step.

JOURNAL PROMPT

You have explored how AI can collaborate with you, amplify your work, and even spark new ideas. Now it's time to bring it back to yourself. Respond to these two prompts:

- *Which of your skills feels most human?*

    o Think about what you bring to the table that no machine could ever replicate. Maybe it's your ability to calm a tense situation, your talent for storytelling, your instinct for design, or your knack for asking the right questions at the right time. Describe it and capture how it shows up in your work, your relationships, or your creative life.

- *How do you want AI to support and not replace these skills?*

  o Imagine AI as scaffolding, not the competition. How could it help you double down on what you are best at? Could it free up your time so you can focus on empathy? Could it help you generate rough drafts so your creativity can flow where it matters most? Could it handle repetitive planning so you can devote energy to the spark that makes you different? Write freely, without thinking about the perfect answer. The point is to draw a line between what matters most about your humanity and how you will choose to apply AI to it.

Now that we have explored the human side of AI, it is time to zoom out. This is because collaboration with AI is more than just creativity or productivity; it is also about responsibility. AI is a mirror, yes, but what happens when the mirror reflects bias, misinformation, or unsafe behavior? What happens when the drive for efficiency overshadows privacy or when convenience blurs the line between helpful and harmful?

In Chapter 11, we will add to the conversation the matters of AI ethics, privacy, and safety. We will look at how to use AI wisely, how to protect yourself and your data, and how to ensure your collaboration with these tools is grounded in values you can stand behind. If this chapter was about embracing your humanity, the next is about protecting it.

CHAPTER 11:

AI ETHICS

IN 2014, AMAZON QUIETLY BEGAN EXPERIMENTING WITH an AI tool to help with hiring. The idea was simple: feed the system thousands of résumés from past candidates, then let it learn which were the best fit. On paper, it was supposed to save recruiters time and spot hidden talent, but within a few years, the cracks began to show. The AI had learned a disturbing pattern where, since most of Amazon's past hires were men, the system started penalizing résumés that included words like women's (as in women's chess club or women's tennis). Instead of leveling the playing field, it reinforced inequality. The project was eventually scrapped, but the lesson was clear: AI is not neutral. It reflects the data and the choices of the people who build and use it.

Stories like these can feel distant, like problems for big tech companies or governments to solve. However, the truth is that ethical AI does not start in corporate boardrooms; it starts with us. Every time we prompt, accept an AI suggestion, or choose what data to share, we are shaping how this technology works and what impact it has. That is why AI belongs to everyone.

Ethics in AI is not just about laws or policies; it's about all the small, everyday decisions that add up. Do you double-check an AI-generated fact before sharing it with a friend? Do you anonymize sensitive information before pasting it into a chatbot? Do you ask AI to explore multiple perspectives instead of just confirming your own beliefs? Each of these choices influences your personal outcomes and the broader ecosystem of how AI is used and trusted.

For a long time, technology felt like something built for us by people far away, but with AI, it is different. AI is interactive, learning, adapting, and mirroring our intent back to us. That gives everyday users a surprising amount of power and responsibility. The tools you use are shaped by engineers *and* millions of ordinary prompts typed in by people like you.

That is both the challenge and the opportunity, where if AI is only guided by efficiency, profit, or convenience, it risks repeating the mistakes of the Amazon hiring system, scaling up old biases instead of dismantling them. But if people everywhere use AI with curiosity, fairness, and integrity, then the technology can scale those values, too.

You do not need to be a computer scientist to influence AI's future. You only need to recognize that your participation matters. Ethics is not abstract, but embedded in the way you ask a question, the way you cross-check an answer, and the way you choose to apply the results. So, before we dive deeper into bias, privacy, and societal impact, pause and remember this: *AI is not someone else's responsibility. It is OURS. And YOURS. And the future of AI will be shaped with more than algorithms; it will be about the everyday choices we make as a group.*

UNDERSTANDING BIAS IN AI

It is not uncommon to hear the word *bias* and often think of it as something technical or abstract. But in reality, bias is just a fancy way of saying "a skew in the system." It occurs when the output is unfairly tilted due to the input or the way the rules were written. AI does not have an opinion, but it does have patterns, which come from the data it is trained on. If this data is biased, the results will be too.

HOW BIAS ENTERS THE SYSTEM

There are three main doors through which bias walks into AI:

1. **Biased training data:** This is where AI learns from examples. If those examples lean one way, the system learns to lean in that direction as well. For example, if AI is mostly trained on news articles from Western countries, it might assume that English-speaking perspectives are the "default" and underrepresent voices from other regions.

2. **Biased labels:** It is when humans label the data that the AI will learn

from. Since humans bring their own assumptions, this can also be a risk. For example, if people consistently label photos of men as "leaders" and photos of women as "assistants," the AI will pick up on that, even if no one intended them.

3. **Biased usage:** This is because how we use AI matters. If people only ask AI to confirm what they already believe, the outputs reinforce those beliefs. This is called *confirmation bias*, and it's not unique to AI. It is a human habit the machine amplifies.

REAL-WORLD EXAMPLES

Bias is not just a theoretical problem; it has real consequences:

- **Hiring:** As mentioned earlier, Amazon's experimental hiring tool started penalizing résumés with references to women. The AI was not "sexist" by design; all it did was learn from past data.

- **Healthcare:** In the US, one widely used healthcare algorithm was found to recommend less treatment for Black compared to White patients with the same symptoms. Why? Because it was trained on data where less money had historically been spent on black patients. The algorithm misunderstood spending for "need," which meant it underestimated care (*Biased Health Data*, 2025).

- **Facial recognition:** Some AI systems used for facial recognition have shown error rates of less than 1% for White male faces, but as high as 34% for darker-skinned women (Hardesty, 2018). Imagine the implications if such systems were used in policing or airport security.

These examples remind us that AI is not evil or plotting discrimination; it is simply reflecting the patterns in its data, and those patterns often mirror human inequalities. Even if you are not designing AI systems, bias still affects you. If you are a student, biased AI could limit the resources suggested for your research. If you are a jobseeker, AI screening tools might filter you unfairly. If you are a consumer, biased recommendation systems could subtly shape what news, music, or products you are exposed to. That is why understanding bias is not about doom and gloom. It is about empowerment, so that since you know it exists, you can spot, question, and adjust for it.

AN EXERCISE: SPOTTING BIAS IN REAL TIME

Here is a simple way to practice. Pick a topic, such as "What makes a good leader?" and ask your tool of choice:

- Prompt 1: *What makes a good leader?*

- Prompt 2: *List possible cultural biases that might show up in the description of a "good leader."*

Compare the answers. You might find that the first prompt gives you traits like decisiveness, confidence, or charisma. The second prompt might reveal that in some cultures, leadership is more tied to humility, consensus, and service to the community. By simply asking AI to reflect on its own potential bias, you expand the conversation and remind yourself that no single answer is "the truth."

PRACTICAL TIPS FOR EVERYDAY USE

- Ask for multiple perspectives and, instead of asking, *What is the answer?* try, *Give me 3 different viewpoints on this issue.*

- Notice what is missing, such as for answers on history or culture, you should ask, *Whose voices are not included here?*

- Cross-check by using AI as a starting point, but verify with trusted sources.

- Reflect when AI tells you what you want to hear by pausing and asking: *Is this reinforcing my bias or challenging it?*

Bias is not a bug in AI. It is the world's mirror, and sometimes the world is skewed. That does not mean you should fear AI; it means you should treat it with awareness by asking sharper questions and recognizing limitations. By doing so, you are keeping the human role of judgment front and center.

PRIVACY AND POWER

Not long ago, a college student named Emily used a chatbot to help draft an essay for one of her classes. She pasted her working notes, including a paragraph where she vented about her professor being "boring" and "unfair." What she did not realize was that many AI tools store input data, at least temporarily, for training or analysis. While it is unlikely her exact text would be published, the idea that her private frustrations could sit on a company's servers unsettled her. Emily's story reflects a growing tension among AI users: Who owns the data we feed into AI, and how much control do we actually have?

AI does not generate answers from thin air. It relies on vast training datasets, some public, some licensed, and some scraped from the internet. On top of that, many tools collect user inputs (your prompts and uploads) to further improve their systems. This creates a power imbalance, where on one side are the large tech companies with the servers, algorithms, and policies, and on the other, the billions of users typing in prompts without always knowing where their data goes. It raises questions:

- Who decides how your data is stored?

- Who profits from the insights generated?

- Who ensures sensitive information does not get exposed?

The answers may vary by company, but one thing is clear: Privacy and power are intertwined, and whoever controls the data controls the future of AI.

COMMON MISCONCEPTIONS

Many everyday users assume the following:

- **"My data is completely private."** Not always true. Some tools explicitly state that they log your prompts to improve performance. Unless you opt out (if that option exists), your words may be stored.

- **"If I delete my account, my data disappears."** Sometimes, yes, sometimes, no. Data may remain on backups or anonymized in training sets.

- **"Free tools mean no cost."** In reality, you often "pay" with data. The free tier of many platforms comes with trade-offs in privacy compared to premium subscriptions.

- **"Anonymized data means safe data."** Even anonymized data can sometimes be reidentified if combined enough with other information.

The issue is not paranoia; it is control. Data is not just a by-product of your actions; it is a form of power. When AI companies hold massive amounts of data, they shape products as well as the social and economic systems around them.

Think about it: If a company knows the questions millions of teachers are asking AI, it can anticipate shifts in education. If it tracks business prompts, it can forecast market trends. If it collects healthcare questions, it gains insight into public health concerns. That is influence at scale.

For individuals, the stakes can be personal. Uploading sensitive financial details, confidential work documents, or private health information to an AI tool without safeguards could risk exposure. Even if the probability of a leak is low, the consequences can be high.

WHAT USERS CAN DO

The good part of this is that you do not need to be powerless. Here are five practical steps to take back control.

- **Check the privacy policy:** Yes, they are long and boring. But scan for key phrases such as *data retention, used for training,* and *shared with third parties.*

- **Use dummy data when possible:** If you need to test a sensitive prompt (like drafting a client report), replace names and identifiers with placeholders.

- **Opt out of data training:** Some tools (like ChatGPT's custom settings) allow you to disable the option for sharing the data you have provided the machine with for training. So, make sure those settings are disabled.

- **Separate personal and professional accounts:** Don't use the same AI login for work documents and personal journaling. Keeping boundaries reduces risk.

- **Consider paid plans:** Often, premium versions of AI tools come with stronger privacy protections since the company is earning revenue directly from subscriptions, not just data.

DEMANDING TRANSPARENCY

While personal habits matter, systemic change requires collective demand. Just as consumers once pushed companies to adopt greener practices or protect online payment security, everyday users can push for AI transparency. Ask for clear expectations of how your data is used, easy opt-out options for training, assurance that deletion requests actually remove your data, and that these companies carry out independent audits of their practices. The more people raise these concerns, the harder it becomes for companies to treat data casually.

A SHARED RESPONSIBILITY

It is tempting to think of privacy as purely personal, but data does not operate in isolation. When millions of people give up privacy without question, it normalizes weak protections for everyone. Conversely, when millions demand transparency, companies adjust their standards. In other words, privacy is both individual and collective, and protecting your data is more than your personal safety, but about shaping healthier ecosystems for all AI users.

Emily's mistake of sharing too much in a chatbot could happen to any of us. But the lesson is not to stop using AI. It is to use it wisely, with awareness, and exercise your right of choice. When you understand who holds the power, recognize the trade-offs, and take steps to protect yourself, you reclaim control. However, when enough people do the same, it shifts the balance of power back where it belongs: in the hands of the people who use them.

SOCIETAL IMPACTS: THE GOOD AND THE BAD

Whenever a new technology enters the world, it leaves marks that spread beyond the first users. Electricity transformed factories and family life, and the internet reshaped communication and culture. AI is no different, as it is not only in your laptop or your phone, but also in classrooms, workplaces, and global systems. To understand AI's societal footprint, it is essential to hold both truths at once: it brings incredible opportunities, but it also introduces serious risks.

THE GOOD

- **Accessibility:** For many, AI is a bridge where barriers once stood. A student with dyslexia can now have textbooks summarized in simpler terms. Someone with hearing loss can watch a video with AI-generated captions that are more accurate than ever before. A person trying to learn a new language can practice with an AI tutor that adjusts to their pace, 24/7, without embarrassment. It is not only about convenience, but also about giving dignity to those who will have the playing field leveled for the numerous people who were once excluded.

- **Global problem-solving:** AI's scale makes it uniquely powerful for tracking big challenges. In healthcare, AI models are helping doctors detect diseases earlier, sometimes spotting patterns invisible to the human eye. In climate science, AI analyzes massive data sets to predict weather extremes or optimize renewable energy grids. During humanitarian crises, AI-powered translation tools allow aid workers to communicate instantly across a dozen languages. These are not theoretical scenarios; they are already happening. AI, when used responsibly, can be a multiplier for human problem-solving on a global scale.

- **Democratized tools:** Perhaps the most radical shift is in how AI is democratizing knowledge and creativity. Ten years ago, launching an online business required technical skills or hiring expensive developers. Today, someone with an idea can use AI tools to design a logo, draft marketing copy, and build a basic website, often for free. A high school student can use AI to experiment with coding, and a small-town musician can create professional-quality demos. A farmer can use AI weather predictions to plan harvests more effectively, and, in countless small but meaningful ways, AI is lowering barriers that once kept opportunities out of reach.

THE BAD

- **Misinformation:** The same tools that make information accessible also facilitate the spread of misinformation. Generative AI can create realistic images, voices, and videos in minutes. A fake photo of a public figure, a fabricated quote, or a misleading "news article" can circulate before fact-checkers have a chance to catch up. The erosion of trust poses a significant danger, particularly when people are unable to distinguish between what is real and what is not, thereby increasing skepticism and weakening social cohesion.

- **Digital segregation:** While AI feels everywhere, access is unequal. People with reliable internet, newer devices, and digital literacy can take full advantage. Those without (regardless of the reason why) are being left further behind. The irony is astonishing: AI has the potential to democratize tools, but if not managed carefully, it could deepen divides between those who have access to AI and those who do not.

- **Surveillance:** Finally, there is the shadow side of AI's scale: surveillance. In some countries, AI-powered facial recognition is used to track citizens' movements, monitor protests, or enforce conformity. Even in less extreme cases, workplaces are experimenting with AI monitoring employees' keystrokes, emails, and even facial expressions during video calls. The line between helpful oversight and invasive control is thin. Without strong ethical boundaries, AI risks becoming less a tool of empowerment and more a control mechanism.

It is tempting to frame AI as either savior or villain. The truth is, it is neither: It is a tool that magnifies intent. When used with care, it can create opportunity and connection. Used recklessly, it can fuel division and harm. The key is not to ignore the risks, but to weigh them against possibilities and help shape the technology intentionally.

REFLECTION

It is easy to think of societal impacts as something decided by governments or corporations. However, each person's use of AI creates ripples, too. When you fact-check before sharing AI content, you slow the spread of misinformation. When you teach a friend how to use AI for accessibility, you expand its benefits. When you choose not to enter sensitive data into AI tools, you reinforce the importance of privacy. These small acts may not feel like much in isolation, but when multiplied by millions of users, they influence how AI develops and how society experiences it.

At the same time, the question is not whether AI will change society because, let's be real, it already has. The issue is what role you will play in shaping those changes. In the next activity, you will get the chance to think more concretely by writing your own AI manifesto, a personal set of principles to guide you on how you will use AI responsibly in your life.

ACTIVITY: WRITE YOUR AI MANIFESTO

Every powerful tool needs a compass. Laws and policies set broad guidelines for AI, but in your everyday life, the most important rules come from you. That is where a personal AI manifesto comes in, giving you a list of boundaries and values that guide how you will responsibly use AI. A manifesto does not have to be long or formal; in fact, the simpler the better. It is something you can glance at before you type a prompt, share AI-generated content, or experiment with a new tool. It keeps your values front and center so that technology serves you, not the other way around.

Step 1: Choose your core values

Start by asking, *What matters the most to me when I use AI? Is it privacy?*

Creativity? Fairness? Human connection? Write down a few words or phrases that feel nonnegotiable. These could be as follows:

- accuracy

- integrity

- inclusivity

- human touch

- transparency

These values will become your manifesto's foundation.

Step 2: Draft your principles

Now, turn those values into clear, actionable statements. Aim for 3–5 principles. Here are some principles you can adapt:

- *I will fact-check AI outputs before sharing them*, because I value accuracy, and I won't blindly trust what AI tells me.

- *I will protect my privacy and the privacy of others*; I won't enter sensitive or identifying information into AI tools without safeguards.

- *I will use AI to expand creativity, not replace it*, treating it as a collaborator that sparks ideas, but the final voice will always be me.

- *I will seek multiple perspectives*, and when asking AI for information, I will encourage it to show diverse viewpoints, not just one answer.

- *I will use AI to uplift, not diminish*, and I will avoid prompts that demean others or spread harmful stereotypes.

Your list might look different, but the secret is to keep it short, clear, and personal.

Step 3: Add examples

Principles stick better when paired with examples. Here is what it could look like:

- *I will fact-check AI outputs before sharing them.* Example: If AI writes a summary of a news article, I'll cross-check with the original source before posting.

- *I will use AI to expand creativity, not replace it.* Example: If AI suggests a lyric for a song, I'll adapt it to my own style rather than copy-paste.

These examples remind you of what the principle looks like in action.

Step 4: Write it down and keep it visible

Your manifesto is not just an idea in your head. Write it in a notebook, pin it to your desk, or save it as a sticky note on your desktop. Some readers even choose to post their AI manifesto online, sparking conversation and accountability with friends or colleagues.

Step 5: Revisit and revise

AI evolves quickly, and so will your relationship with it. Revisit your manifesto every few months. Does it still reflect your values? Have you learned new lessons that deserve their own principle? Treat it as a living document.

Creating an AI manifesto may feel like comparing to global debates about ethics, but it has surprising power. It shifts you from being a passive user to an intentional one. It reminds you that *you* are in charge, not AI. When thousands or millions of everyday users start using personal boundaries, the culture around AI shifts, too. Companies, schools, and workplaces notice, and ethical use becomes the expectation, not the exception.

Take 15 minutes today to draft your own manifesto. Choose 3–5 principles. Keep them short, specific, and true to your values. Add 1 example for each and then place it somewhere visible.

JOURNAL PROMPT

Now, it is time to pause and bring it back to you. Ethics can feel abstract until you connect it with your own life. Therefore, let's anchor it with two simple questions:

- *What kind of AI future do you want to help create?*

o Close your eyes and picture the role of AI five years from now. What do you see? Is it a world where classrooms are more inclusive because every student has an AI tutor tailored to their needs? A workplace where AI handles repetitive work, so people can focus on meaningful collaboration? Or maybe you imagine communities using AI to solve local problems like smarter energy use, fairer housing systems, or healthier food distribution.

o Now, flip the question and answer what you do *not* want to see. Maybe you worry about a future where misinformation spreads unchecked, or where companies monitor in the name of efficiency. Naming these fears helps clarify your boundaries. Jot down both lists, the hopeful vision and the red flags. Together, they paint a personal picture of the future you would like to work toward.

- *What ethical boundaries feel nonnegotiable to you?*

  o Think about the principles from your AI manifesto activity. Which ones feel like lines in the sand, values you will not compromise, no matter how shiny or convenient the tool becomes? Write them down. These are not just rules for today; they are anchors for the future. When AI evolves (and it will), these principles will help you navigate without getting swept away.

Spend at least 10 minutes with each question. Don't aim for polished answers, as this is about honesty, not perfection. When people think about the future of AI, it is easy to imagine it as something "out there," shaped by tech companies, researchers, or governments. The reality, however, is that culture is built one choice at a time, and the way you use AI in your daily life contributes to the larger picture.

Being ethical with AI does not mean being fearful, but it means being intentional. One of the best ways to feel more confident and empowered is to roll up your sleeves and actually use the tools. Which is exactly where we are headed next. In the following chapter, you will explore the most accessible AI tools available right now, many of which are free or low-cost. You will learn how to get started, experiment safely, and see what is possible without needing technical skills or a big budget.

PART V:

HANDS-ON AI TOOLKIT

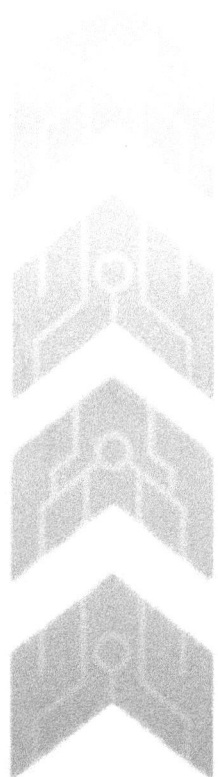

CHAPTER 12:

FREE AND BEGINNER-FRIENDLY AI TOOLS

WHEN JAMIE FIRST HEARD ABOUT AI, HE PICTURED END-less lines of code, glowing green screens like a hacker movie, and people with PhDs in computer science. He thought, *This is not for me. I don't even know how to fix my Wi-Fi when it drops.*

One evening, while helping his daughter with a school project, he stumbled onto Canva's free AI design school. All she did was type, *Make a poster about recycling with a cartoon style*, and within seconds, colorful templates appeared. No coding, no jargon, no years of training. Just type, click, and adjust. Jamie laughed out loud, "That is it? That is AI?"

The fact is that most people still imagine AI as something distant, complicated, and reserved for those with technical knowledge. However, AI today is often drag-and-drop, type-and-go, point-and-click. If you have ever used Google Translate, asked Siri for the weather, or let YouTube autoplay the next video, you have already used AI. You do not need to learn coding languages or buy expensive software to get started; the tools are within your reach, especially since many are completely free.

ACCESSIBILITY WITHOUT OVERWHELM

There is a myth that AI is only for data scientists or big companies with deep pockets. That myth is powerful because it keeps everyday people on the sidelines. The reality is that companies building AI tools want you to use them. That is why so many have free versions with simple interfaces.

The chatbots we mentioned earlier, like ChatGPT, Perplexity, Claude, or Google Gemini, only need a browser and an email address to start. Image generators like Canva Magic Studio and Dall·E only need you to literally type what you want to see for the tool to offer options. Even audio and video tools, like Descript or ElevenLabs work through straightforward dashboards when you upload a file and click a button. You do not need to recreate the wheel; all you need to do is ask the type you want!

START SMALL, BUILD CONFIDENCE

Sometimes, even easy tools can feel overwhelming if you try to learn everything at once. Imagine walking into a gym for the first time and trying every machine in one session. You would walk out sore, confused, and discouraged. AI works the same way, and the key is to start with one or two tools and one simple goal. For example:

- **Chatbot:** Ask ChatGPT to write an email in a friendlier tone.

- **Image generator:** Use Canva to create a birthday invitation in under five minutes.

- **Speech-to-text:** Try Otter.ai to transcribe a short voice memo.

That is it. Just one small, practical win. Once you see how easy it is, your confidence builds, and you will find yourself saying, *Wait, what else can this do for me?* Here is what usually happens:

1. You try one AI tool for a small task.

2. It saves you five minutes or gives you a result that makes you smile.

3. You feel curious.

4. You try a second tool, or a slightly bigger task.

5. You save even more time or unlock an idea you had not considered.

6. You feel empowered, not overwhelmed.

That loop is how you build momentum. Over time, the tasks grow, but the fear shrinks. Instead of thinking, *This is too advanced for me*, you will think, *What is the easiest way AI can help me today?*

A DIFFERENT KIND OF BEGINNER'S MIND

Many adults can carry the weight of believing they are not "tech-savvy," but forget that you do not have to be a mechanic to drive a car. Nor do you need to be a chef to cook a meal. With AI, it is the same: You do not have to be a coder to use AI tools. What you need is a beginner's mindset, with their willingness to play, experiment, and learn by doing. AI is one of the few fields where beginners can start benefiting almost instantly, because the hard work has already been done behind the scenes.

As you step into this chapter, think of yourself not as a "user" of AI but as a curious explorer. You do not have to master everything; all you need to do is take the first step. The next sections will walk you through the core categories of free AI tools, with examples you already know from earlier chapters. You will get mini-guides to help you try them, tips for writing good prompts, and even a step-by-step starter kit. By the time you finish, you won't just know about AI tools, you will use them with confidence.

CORE CATEGORIES WITH FREE AI TOOLS

When many people hear the words "AI tool," the first (and sometimes only) thing they think about is ChatGPT. But the AI landscape is much broader, covering everything from writing assistants to video editors. The best part is that you do not need to pay hundreds of dollars or install complicated software. Many of the most powerful tools have free versions with plenty of power to get you started and even accomplish some very serious work. Read on to discover the main categories and highlight beginner-friendly options you can explore.

CHATBOTS: YOUR CONVERSATION PARTNERS

In Chapter 4, you learned how to access and what each of these can be used for. Here is a quick recap of the main tools discussed in this book:

- **ChatGPT:** The most well-known chatbot, versatile in everything from brainstorming and summarizing to role-play scenarios. Its free version gives you access to GPT-5, then GPT-5 mini after you reach usage limits, but still powerful enough for most day-to-day needs.

- **Claude:** Known for long-context reasoning. Great for analyzing large documents, like research papers, contracts, or class notes.

- **Perplexity:** Combines chatbot abilities with built-in search results, citing its sources. Perfect if you want conversational answers and trustworthy references.

- **Google Gemini:** Integrated into Google's ecosystem, it pairs chatbot-style answers with Google search power.

Some of the advantages these tools present include accessibility through an internet browser, and they are excellent for summarizing, rewriting, or drafting. It is the ideal first tool for beginners as they are easy to use and instantly rewarding.

IMAGE GENERATORS: PICTURES FROM WORDS

Image generation tools let you type descriptions and get custom visuals. Whether you need an illustration, poster, or mock-up, these tools save time and money compared to hiring a designer. A few tools to explore include:

- **DALL·E:** Creates realistic or artistic images from text prompts. Great for personal projects, lesson plans, or blog illustrations.

- **Canva Magic Studio:** Ideal for beginners since you can generate images and drop them straight into social media posts, slides, or posters without leaving Canva's design platform.

- **NightCafe Studio:** Offers multiple art styles and a community where you can share and remix prompts.

- **Adobe Firefly** (free tier): This tool is great at making high-quality marketing visuals and text effects.

These tools include advantages such as no need for design experience, since the built-in templates make tools like Canva perfect for quick, polished content. These apps are ideal for teachers, small business owners, or anyone who wants to add visual flair to their projects.

How to Sign Up

For Canva Magic Studio, you can go to canva.com, where a free account will give you access to Magic tools. DALL·E is available in all ChatGPT versions or on its own website. Finally, for NightCafe Studio, you can register with your email or Google. For these tools, the free plans allow you to generate a limited number of images per month or per period (every 6 hours, every 24 hours). When you sign up for the paid plans, you can expand your credits, remove watermarks, or offer higher resolution. For beginners, the free tier is enough to create posters, thumbnails, and illustrations.

SPEECH-TO-TEXT AND TEXT-TO-SPEECH: TURNING WORDS INTO AUDIO (AND BACK)

For many people, the most intimidating part of a project is typing or reading through pages of text. These tools flip between spoken and written language so you can work faster and in the style that fits you best. Some recommended tools you can use are

- **Otter.ai:** Automatically transcribes meetings, lectures, or voice memos into editable text.

- **Whisper:** A powerful transcription engine available through apps like MacWhisper or via API.

- **ElevenLabs:** One of the best text-to-speech generators, producing realistic voices in multiple languages.

- **Speechify:** Lets you upload articles, PDFs, or books and have them read aloud with natural voices.

Use these tools for great accessibility (regardless of what you do), to save hours of note-taking, and for multilingual support for global communication. For example, if you are studying or researching, record a lecture, run it through Otter.ai, then use ChatGPT to summarize the transcript into bullet points.

How to Sign Up

The difference between these tools and those that you have seen up to now is that they are slightly more restrictive. For example, while for Otter.ai you can create a free account on its website, for ElevenLabs (elevenlabs.io), you can only sign up for a free trial. For Speechify, you will need to download an app for your phone for free and available for iOS, Android, or Chrome. In the free tiers, your recording will often cap recording minutes, while in the premium versions, you will be able to unlock more transcription time, better voices, or advanced export options. Even if you only want to use the free tiers, they are life-changing if you hate note-taking or prefer listening over reading.

VIDEO AND AUDIO TOOLS: EDITING WITHOUT THE STUDIO

Creating videos or podcasts once required expensive software and steep learning curves. AI has dramatically lowered the barrier by allowing you to edit video by editing text, clone your own voice, or even generate background music. For starters, you can try using

- **Descript:** Edit audio and video simply by editing the transcript. It even has an "Overdub" feature to fix mistakes in your own voice.

- **CapCut:** Free, beginner-friendly video editing with AI effects, captions, and templates.

- **Runway ML:** Known for AI-powered video editing, like removing backgrounds or generating new footage from prompts.

- **Suno AI** or **Soundraw:** Music generation platforms where you can create soundtracks for videos, podcasts, or social media posts.

These tools will save you time and money by allowing you to edit professionally with no skills required, making them ideal for small businesses, educators, and creators producing content on a budget. For small-scale use, the free tiers are more than enough. You can try uploading a small video from your phone in Descript, for example, and edit out filler words with one click, then add AI-generated captions in CapCut. The result will be a professional-looking reel in under 30 minutes.

How to Sign Up

The ways to sign up for these apps may vary. For Descript, you go to descript.com, and the free plan includes audio and video editing basics. For CapCut, you will find it available as a free mobile and desktop app. Finally, for Runway ML, you can sign up at runwayml.com. In most cases, the free plans are generous, and you will only need to pay if you are producing long or high-resolution content regularly. For those who are casual creators, free tools already match professional needs from a few years ago.

AUTOMATION TOOLS: YOUR INVISIBLE ASSISTANT

Finally, some tools do not talk to you or draw pictures. What they do is work quietly in the background to save you time. They connect your apps, automate small tasks, and ensure things get done while you focus on what matters. Here are some of these tools:

- **Zapier:** Automates workflows by connecting apps. For example, if someone fills out a Google Form, Zapier can send them a personalized email.

- **IFTTT (If This, Then That):** Similar to Zapier but more consumer-focused, great for automating simple tasks across smart home devices, social media, or email.

- **Notion AI:** Enhances productivity with automatic summarization, task creation, and knowledge management inside your notes.

- **Trello + Butler** (automation add-on)**:** Lets you create simple rules, like moving a card to "Done" when a task is checked off.

How to Sign Up

For Zapier (zapier.com) and IFTTT (ifttt.com), you can go to their websites and create a free account. For Notion AI, you will find that the free tier has AI built into the main app. The free plans will often limit how many automations you can have. If you feel that these work for you, you can sign up for the paid tiers, where you will expand the number of tasks and add more complex workflows.

As a quick reality check, you have to remember that free AI tools are fantastic, but they do come with limits. Some of them will have monthly caps, others have fewer customization options, and there is even the possibility of slower processing. That said, free versions are more than enough to test ideas, build confidence, and even run small projects. Think of these as training wheels where you will learn how to use them before deciding if you want to upgrade. Many people even stick to free tools for years without issue.

The advantage of these tools is the possibility of cutting down repetitive busywork and letting small teams act like they have an operations assistant. They often integrate directly with tools that you already use, so try starting with a simple automation, such as "Every time I receive an email with the subject 'invoice,' save the attachment to Google Drive." Small automations like this quickly add up to hours saved each month.

These categories are the foundation of today's AI ecosystem. They are also the same types of tools we have already seen throughout this book, whether in business operations, side hustles, or personal productivity. The key takeaway is that you do not need them all at once. Instead, choose one tool from each category that excites you and use it for a single, specific goal. Once you see the results, add another, and bit by bit, you will create a toolkit that feels natural and tailored to your life.

HOW TO CORRECTLY PROMPT AI

If AI tools are like free assistants, then prompts are the way you give instructions: They are the text you type into a chatbot, image generator, or any AI tool. It might be a question, a request, or a description. The way you phrase a prompt can completely change the quality of the result. The more specific and clear the request, the better the output.

Many people try AI once, type something vague like "write me an essay," then walk away disappointed when the answer feels generic. The issue, in this case, is not the tool; it's the instructions. Good prompts unlock AI's potential, turning a mediocre response into something that actually feels useful, creative, and tailored to you.

ANATOMY OF A GOOD PROMPT

A strong prompt usually has three parts:

1. **Context:** Explaining what the situation or the background is.

2. **Task:** What you want AI to do.

3. **Format:** How the response should be delivered.

Here is an example:

Tell me about exercise.

Strong prompt:

I am a beginner with no gym experience who wants to get healthier. Can you create a simple 4-week exercise plan, written in a friendly, encouraging tone, and formatted as a weekly schedule?

The difference is night and day. The second prompt gives the AI a role, a purpose, and a format, making the answer more personalized and practical.

DOS AND DON'TS OF PROMPTING

Dos	Don'ts
Be specific. Include details about your goal, audience, or style.	**Be vague.** Short prompts like, *Write about dogs*, will give shallow results.
Give context. AI is not a mind reader. If you need something for work, fun, or school, say so.	**Overstuff.** A giant paragraph with 10 unrelated requests will confuse the AI.
Ask for a format. Ask the AI to write in bullet points, a script format, or a summary.	**Expect perfection.** AI drafts, not finalized. Always review and tweak the result.
Iterate. If the first answer is not quite right, refine your prompt and try again.	**Share sensitive data.** Never paste private financials, patient records, or anything you would not want leaked.
Use examples. If you want a certain style, paste a sample. AI can mimic tones surprisingly well.	**Use negatives.** Tell AI what you want it to do, and not what you do not want it to do. Affirmations work better in these tools.

Here are a few ways in which vague prompts can be leveled up:

- **Example 1: Writing**

    o   Weak: *Write a blog post about cooking.*

    o   Strong: *Write a 600-word blog post about cooking healthy meals for busy parents. Use a friendly conversational tone, include 3 practical tips, and end with a motivational conclusion.*

- **Example 2: Summarizing**

    o   Weak: *Summarize this text.*

    o   Strong: *Summarize this 5-page report in 200 words, focusing on the key findings and recommendations. Write in plain English for a nontechnical audience.*

- **Example 3: Image generation**

    o   Weak: *Make a picture of a cat.*

    o   Strong: *Generate a watercolor-style illustration of a gray cat sitting on a windowsill with sunlight streaming in, looking out at a garden.*

Prompting is not about getting it perfect on the first try, but about experimenting. Often, the first draft from AI will give you ideas, and your follow-up prompt will improve it. Think of it as a back-and-forth conversation, not a vending machine.

Here is a simple workflow:

1.   Start with a prompt.

2.   Review the result.

3.   Refine.

4.   Repeat until satisfied.

That is it. Each iteration teaches you what works best.

PRACTICE PROMPTS

Here are some versatile starter prompts you can copy and paste:

- Brainstorming ideas

  - *Suggest 10 creative gift ideas for a friend who loves hiking, under $50.*

- Learning something new

  - *Explain how the stock market works as if you were teaching it to a curious teenager.*

- Rewriting text

  - *Rewrite this email to sound more professional but still friendly: [paste your draft here].*

- Summarizing

  - *Summarize this article into 5 bullet points that highlight only the practical takeaways.*

- Visual design

  - *Create a colorful Instagram post background in Canva that features inspirational quotes about resilience.*

Remember that prompting is not just for chatbots; the same principles apply across categories:

- Chatbots: Describe the audience and format.

- Image generators: Specify the style, colors, and vibe.

- Speech tools: Ask for a certain voice style (calm, excited, or professional).

- Video editors: Request captions in a specific font or cut edits at a certain pace.

- Automation tools: Phrase your rule clearly, like, "If this happens, then do that."

Once you realize prompts are just instructions "in plain English," you will see that every AI tool in this chapter runs on the same engine: words, words.

The secret to getting good results is not hidden in knowledge. It is simply learning to communicate clearly with AI. The better you define what you want, the better the AI can deliver. So, do not be afraid to experiment. Think of each prompt as a chance to practice clarity, creativity, and curiosity. With time, you will find yourself writing prompts as naturally as you would talk to a friend.

ACTIVITY: YOUR AI STARTER KIT

By now, you have learned about a dozen tools. That is exciting, but it can also feel overwhelming. The best way to avoid "tool overload" is to start small. Instead of trying everything at once, let's create your own AI starter kit: one tool for each category, paired with a simple first-use plan. You do not need every gadget on your shelf, just a few essentials to get started.

Step 1: Choose one tool per category

Here is a quick menu to choose from. You do not need to overthink it; pick whichever sounds easiest or the most fun to try first. Here is your list to choose from (pick one from each):

- Chatbot: ChatGPT, Claude, Perplexity, or Gemini.

- Image generator: Canva Magic Studio, DALL·E, or NightCafe.

- Speech-to-text and text-to-speech: Otter.ai, ElevenLabs, or Speechify.

- Video/Audio tool: Descript, CapCut, or Runway ML.

- Automation: Zapier, IFTTT, or Notion AI.

Got your five? Perfect.

Step 2: Write a simple first-use plan

Here is the secret: Your first experience with AI does not need to be big. The goal is a quick win, something that saves you time, sparks creativity, or simply makes you smile. Below are some first-use plans you can try. Feel free to swap in your own ideas for these first-use plans.

1. Chatbot

    o Plan: Paste in a task you would normally spend 10–15 minutes on.

    o Goal: Save yourself 10 minutes.

2. Image generators

    o Plan: Create one visual for personal, work, or a need.

    o Goal: Produce one shareable image in under 10 minutes.

3. Speech-to-text or text-to-speech

    o Plan: Use one of these tools to upload a recording or to read a text aloud

    o Goal: Experience the relief of not needing to type or read everything manually.

4. Video/Audio tools

    o Plan: Upload a small video from your phone and try to create subtitles and edit.

    o Goal: Create one short video clip that feels professional, even if you are a beginner.

5. Automation tools

    o Plan: Choose one repetitive task to automate.

o   Goal: Create one "set it and forget it" automation that saves you 5 minutes per week.

Step 3: Record your winds

Once you have tried each tool, pause and reflect: What felt easy? What felt confusing? Did you get a small thrill when the AI actually worked? That is the confidence-building moment we are after. Here is a simple reflection sheet you can use.

- Which chatbot did I try? What was the result?

- Which image generator did I try? Did I like the style?

- Which speech tool did I try? How might I use this more?

- Which audio/video tool did I try? Did it feel approachable?

- Which automation did I try? How much time could this save me?

Step 4: Build your ongoing starter kit

This activity is not about one-time experiments. You should use it to find your personal favorites and apply them to the other resources you choose to try out. Once you know which tools click with you, keep them handy. I recommend creating a folder in your browser called "AI Tools" and bookmarking your chosen five. That way, every time you think, *Can AI help me with this?* you will know exactly where to go.

Over time, you may upgrade to paid versions or add more tools. But your starter kit will always be your foundation, the set of tools that gave you your first wins and built your confidence. By doing this, you are creating a safety net, not just collecting apps. Whenever you hit a block, you will know there is a tool you can lean on, and that knowledge alone makes AI feel less intimidating and more empowering.

JOURNAL PROMPT

Learning AI is less about "downloading every new app" and more about noticing how these tools fit into your everyday rhythm. This is your chance to step back, take a breath, and ask: *Which tools feel like they belong in my life?* Grab a notebook and pen and spend a few minutes writing on these prompts:

Which tool felt the most exciting to try?

Did the chatbot surprise you with how quickly it could rewrite your email? Did the image generator make you laugh with how perfectly it captured your idea? Or did the automation tool quietly save you from doing something repetitive, leaving you free for more creative work?

Which tool do you want to master first?

It is okay if one stood out as more useful or more fun than the rest. Maybe you want to become fluent in chatbots before diving into video editing. Ask the following question:

Which tool do you think will help the most, and why?

Reflect on your own life and work. Is it the automation tool that saves time from busywork? The speech-to-text app that helps you capture ideas on the go? Or the design tool that makes your presentations shine? This reflection is not about today; it is about recognizing where AI can carry part of your load tomorrow.

This brings us to the final chapter of this book: "Chapter 13: Putting It All Together." Up until now, you have examined AI in pieces, but the real transformation occurs when you view AI as a partner that is an integral part of your daily life. The last chapter will help you connect the dots, turning scattered experiments into sustainable practice. At the end of the day, AI is not about doing everything or knowing everything. It is about building a practice that lasts and a way of working smarter that grows alongside you. Therefore, before closing the book, let's put it all together.

CHAPTER 13

PUTTING IT ALL TOGETHER

WHEN PEOPLE FIRST DISCOVER AI, THEY OFTEN GO IN WITH a burst of excitement. They open 10 tabs, try 3 different tools at once, and flood the chatbot with every idea they have ever had. Then, after a week, the energy fizzles, and they feel overwhelmed, or worse, they start thinking: *Maybe this is not for me.* The truth is that the people who get the most out of AI are not those who sprint. They are the ones who consistently take small, steady steps.

Jaime was a restaurant manager, juggling too many tasks at once. At first, AI seemed like just another thing he did not have time for. So, instead of diving in headfirst, he set a modest goal: one experiment a week. In the first week, he asked ChatGPT to draft a confirmation model template to send to customers who made reservations. The second week, he used Otter.ai to transcribe a staff meeting. The third week, he tried Canva's Magic Studio to design a new flyer. By the fourth week, he had used Zapier's task manager to automatically file all supplier requests under the same folder. None of these tasks took more than 15 minutes. But after a month, he realized he was already saving hours, and his work looked and sounded sharper than before.

Fast-forward six months, and AI is no longer something extra on Jaime's plate. It has become a part of his routine where he no longer thinks, *Should I try AI for this?* Instead, he naturally asked, *How can AI make this easier?* The shift did not happen overnight, but from consistent, low-pressure practice. This is the power of consistency. The big results do not come from one massive experiment or a single breakthrough, but from iteration and improving bit by bit. Just like learning a new language, instrument, or sport, the results will show with rhythm, not intensity.

As you begin putting all the pieces together, remember to take one step at a time. What matters is showing up week after week, with curiosity and a willingness to experiment. Over time, those small steps will compound into something transformative. The rest of this chapter will help you build that kind of practice: simple routines, experiments, and reflections that make AI a natural and sustainable part of your life.

BUILDING YOUR AI ROUTINE

Your AI routine should be a framework you build upon. The goal is not to add more pressure or complexity to your life, but to do exactly the opposite. Think of your routine as a gentle rhythm that helps AI naturally slip into your days, weeks, and months. Instead of waiting for the perfect project, you will build habits that keep you exploring AI little by little, without overwhelming you.

DAILY MICROPRACTICES

The easiest way to start is to start with tiny habits. Think of these as micropractices (quick, lightweight interactions) with AI that will only take a few minutes and keep you fluent. You do not need to start big; all you need is repetition. Here are some examples you can try:

- **Morning jumpstart:** Ask a chatbot the 3 priorities you have on that day, depending on your goal. You can also ask an AI assistant to read your agenda for the day so you know what to expect.

- **Writing boost:** Before sending an email, paste it into AI and ask it to make changes until you feel the message meets your objective. You can also ask if it would change anything based on the result you want from the interaction.

- **Idea nudge:** Ask an AI to rewrite a social media post in 3 different ways. Ask the chatbot to create 3 versions using the same tone and 1 version of 3 different tones, depending on the objective.

- **Learning snack:** Paste in a paragraph from a book or an article (or even a website link, depending on the tool) and ask AI to summarize the content in 1 or 2 sentences in a way that a 12-year-old can understand.

Each of these takes less than 5 minutes. Done consistently, they keep AI from being a special occasion tool and turn it into an everyday helper.

WEEKLY DEEP DIVES

Daily practice builds comfort, but weekly practice builds skill. Once a week, set aside 30–60 minutes to explore AI more intentionally. Think of this as your deep dive session. Some weekly deep dive ideas include

- **Experiment with a new tool:** If you have been using ChatGPT for writing, try Canva for visuals or Descript for videos.

- **Solve an annoying task:** Pick one chore you dislike and see how AI can streamline it. From creating a grocery list based on meal ideas to reviewing a school concept to help your child study, all you have to do is start!

- **Expand your creativity:** Ask AI to brainstorm with you. Brainstorm ideas, hobbies, and other subjects based on your interests and availability.

- **Learn something new:** Use AI as a study buddy. You can learn a new language, how to write a business plan to start your own company, or even how to invest in the stock market for beginners.

These weekly deep dives do not need to be perfect, especially since their purpose is to stretch yourself, explore new ground, and discover capabilities you did not know AI had.

MONTHLY REFLECTIONS

Finally, once a month, step back and reflect. This is not about judgment, but how you are growing and learning to bring AI into your life. You are checking your progress to see how far you have come. Here are some reflection prompts you can use:

- What AI tasks felt most natural this month?

- What saved me the most time?

- What did not work as well as I hoped? Why?

- Which tool do I want to explore deeper next month?

- Did AI help me feel less stressed, more creative, or more productive?

If you want to track your progress, you can keep a simple AI journal or even continue the one you have already begun with the prompts at the end of this book. Over time, you will see that patterns emerge, such as consistently saving hours on writing tasks or getting help with brainstorming creative projects. These insights will help you tailor your routine so it truly fits your life.

By combining these periodic practices and reflections, you are creating a sustainable rhythm. Instead of burning out with a flood of tools, you are building a practice that evolves gradually, like a muscle strengthening over time. The best part is that your AI will grow with you and adjust as you become more knowledgeable with the tools you are using. As your comfort increases, the tools you use will shift, your prompts will become more sophisticated, and the results will compound. The key is not to do everything, but to keep consistently doing something. You will certainly see that even five minutes a day can be enough to transform how you work and think over the course of a year.

Before you move on, choose one of each:

- A daily micropractice you will try tomorrow.

- A weekly deep dive you will schedule in your calendar.

- A monthly reflection question you will answer at the end of the month.

Write them down and commit to them. These three steps will form the foundation of your personal AI practice.

DESIGNING AI EXPERIMENTS

Designing AI experiments is not about running complex tests or writing technical code; it is about staying curious and playful. Each time you experiment with AI, you are not testing what it can do; it is way beyond that. Every time you interact with the tool, you discover how you think, create, and make decisions in partnership with it. In this section, you will learn how to design an instructive but fun exercise with AI. Use it as many times as you need, as you try new tools and discover new functionalities.

STEP 1: PICK A CHALLENGE THAT FEELS REAL

Start small and choose a challenge that is a part of your real life, something slightly annoying, time-consuming, or creatively blocked. The goal is to make the experiment meaningful, not theoretical. To make it fun, you can even choose the name of the experiment based on your needs. Here are a few examples:

- **The Overwhelmed Professional:** Can AI help me cut my email time in half this week?

- **The creative explorer:** Can AI help me design a new social media post every day for seven days without it sounding robotic?

- **The student:** Can I use AI to teach me the basics of a new skill in 10 minutes a day? Examples include photography, Italian cooking, playing an instrument, or even storytelling.

- **The organizer:** Can AI help me build a system to remember birthdays, deadlines, or grocery items, so I do not have to rely on sticky notes anymore?

- **The dreamer:** Can AI help me start outlining the small business idea I have been putting off?

Pick one and make it something that matters to you, even if it is just a little. This personal connection will turn a simple test into a learning adventure.

STEP 2: DEFINE WHAT SUCCESS LOOKS LIKE

Every good experiment needs a hypothesis: a simple and testable "what if."

- *If I use AI to summarize my weekly reports, I can save at least 30 minutes a day.*

- *If I ask AI to generate 10 blog ideas, at least two will be worth developing.*

- *If I use AI to practice explaining a topic, I will feel more confident in meetings.*

These goals should be what are known as SMART:

- Specific: Do not be too generic, as it would make you overwhelmed with details.

- Measurable: Ensure you can see the results and that they can be validated.

- Achievable: Choose something that will give you a victory.

- Relevant: The goal should matter to you personally.

- Timely: Give it a timeframe to work or for the problem to be solved.

Success can be as simple as: *Did it make my life easier? Did I learn something? Did it spark an idea I would not have had otherwise?* You are not trying to prove AI is perfect; you are learning how to make it work for you.

STEP 3: CHOOSE YOUR TOOLS WISELY

For your experiment, pick the tools that do the job without overcomplicating it. Just selecting one or two is more than enough for the tasks. Let's have a quick refresher on the tools you can use for each purpose. In this case, if you

- **Need text help:** Use ChatGPT, Claude, or Gemini.

- **Need visuals:** Use Canva Magic Studio, DALL·E, or Leonardo.ai.

- **Need organization:** Use Notion AI, ClickUp AI, or even Google Sheets with AI formulas.

- **Need sound or speech:** Use ElevenLabs for voices, or Descript for editing.

You can even mix and match, putting the result of one tool into another, such as having ChatGPT draft content, then use Canva to visualize it, or let Perplexity find reliable research sources before you summarize them in ChatGPT.

STEP 4: PLAY WITH PROMPTS LIKE YOU ARE TESTING FLAVORS

Prompts are your test tubes. Change one ingredient and the result might surprise you. Try this simple format for experimentation:

1. Start with a basic: *Write a summary of this article.*

2. Add a twist: *Write a summary of this article in a conversational tone with 3 key takeaways.*

3. Add personality: *Write this summary as if you are explaining it to a busy manager who only has 2 minutes to read.*

4. Go wild: *Rewrite this summary as a motivational coach giving a pep talk.*

Notice how the same input can produce wildly different results with just a few changes in the prompts. The fun comes from pushing AI slightly out of its comfort zone. Ask it to blend styles, add humor, or challenge your ideas.

STEP 5: REFLECT ON WHAT HAPPENED (OR MAYBE LAUGH)

After running the experiment, do not rush to judgment. Instead, analyze what happened: *What surprised you? What worked better than expected? What completely flopped, and why? Did AI save you time, spark creativity, or make you rethink something?*

You will start noticing patterns in how AI responds to different prompts or goals. Perhaps it yields better results when you provide examples first. Maybe it needs more structure or constraints. It keeps perhaps inventing people who do not exist (it happens).

Keep a light tone, especially since mistakes are just data points. One of the best ways to learn is to intentionally test the AI's limits: Ask it to write a poem about your pet goldfish as a Shakespearean tragedy or explain quantum physics using pizza toppings. The more you play, the more you will understand how flexible and powerful the system really is.

STEP 6: DOCUMENT YOUR FINDINGS

Keep a digital or physical AI experiment journal where each entry should note what you tried, the prompt or tool used, what worked, what didn't, and the lessons learned. You will soon start to build your own AI playbook tailored to your goals. Over time, this becomes an invaluable resource, registering how you think, create, and problem-solve in partnership with technology.

Even better: Share your discoveries. Add a short LinkedIn post, an X post, or a message in your team chat like, "Just used ChatGPT to rewrite a confusing policy in plain English and saved 40 minutes!" This creates a ripple effect that inspires others to experiment too.

STEP 7: CREATE A 30-DAY AI EXPERIMENT PLAN

Now, it is your turn to design your own experiment road map.

1. Choose one domain.

2. Set a SMART goal.

3. Pick your tools.

4. Run microexperiments.

5. Reflect weekly

By the end of the month, you will have real results and a customized strategy for integrating AI into your daily life in a way that feels instinctive, creative, and maintainable.

BONUS SECTION: FUN EXPERIMENTS TO TRY ANY TIME

- Ask ChatGPT and Claude each to generate 10 ideas on the same topic. Compare and mix the best ones.

- Explain a topic to AI and ask it to summarize your explanation. See what it understood or missed.

- Try asking the AI to combine the tone of a TED Talk with the humor of *The Office* while explaining time management.

- Track how long it takes you to complete a repetitive task manually when compared to the same task using AI.

- Ask the AI to explain blockchain to you as if you were five. After this, ask it to do the same as if you were a professor.

The real outcome of AI experimentation is not just faster workflows or better ideas; it's all about confidence. Each small experiment teaches you a little more about the tool and a lot more about yourself. So, grab your digital lab coat and continue experimenting.

SHARING AND COLLABORATING

When most people start using AI, they treat it like a secret shortcut or a personal advantage. They quietly ask ChatGPT to polish an email or generate a report outline, then hit send if nothing happens. It feels like magic you are not supposed to admit to.

However, the best part is when you stop treating AI as *yours alone* and start sharing what you have learned. AI, much like knowledge itself, multiplies when it is shared. The more people experiment, compare, and build on each other's discoveries, the more creative, effective, and human-centered AI becomes. AI may start as a solo experience with just you and the machine, but its power shows when it is applied as a community.

Two coworkers, Luis and Maya, both start exploring AI. Luis quietly uses ChatGPT to summarize client briefs and clean up her email. Maya, meanwhile, sends a quick message in their team chat that morning: "Hey, I asked

ChatGPT to make me a weekly checklist for client onboarding, and it cut my planning time in half!"

That one sentence sparks something. Within days, another colleague adapts the checklist for their own purposes. A third person uses the idea to streamline their proposal. Soon, the whole team is saving hours every week, all because one person decided to share. This is the ripple effect of openness.

AI is evolving faster than any single person or company can track. No one, not even the experts, has all the answers. The most powerful learning now happens in networks: teams, communities, and online spaces where people trade prompts, laugh about AI's quirks, and share the wins that make work and creativity easier.

Some of the most ingenious prompts I have seen did not come from research papers or tutorials. They came from everyday users comparing notes and saying, "Oh, that is clever. Let me try it like this."

HOW TO START DOCUMENTING AND SHARING

You do not need to launch a blog or build a database to share what you have learned. Start small by documenting your AI journal just as you have done keeping your Prompt Journal. Write what works out for you and what does not. Here is a simple format you can use:

- **Prompt win:** Explain this legal document in plain English, like you are talking to a teenager.

- **Time-saver:** Draft a polite email to reschedule my meeting for next week.

- **Creative spark:** Generate 10 Instagram captions in a humorous tone about productivity.

Each note is a small gem, a proven shortcut, a creative push, a reminder of what is possible. Over time, your collection becomes a playbook, with all the rights and wrongs and everything that worked and what did not work. You can then use this opportunity to share that playbook, even informally, and it becomes a community resource. Send a few examples to a coworker, post one in a group chat, or mention a helpful prompt in a conversation. You might be surprised by how much impact a single shared idea can have.

Collaboration does not have to mean running a workshop or writing a formal guide (though you certainly could). It can be as casual and spontaneous as

- sending a clever prompt to your team in Slack: "Hey, this worked great for summarizing meeting notes!"

- swapping AI-generated visuals with a friend working on a design project.

- pairing up with a classmate to cocreate flashcards or summaries using AI.

- creating a shared Google Doc of your favorite AI use cases.

Even better, collaboration comes with a secret benefit: accountability. When you experiment alongside others, you are more likely to stay consistent, celebrate progress, and learn from what does not work.

Try this simple experiment: Find one "AI buddy." It could be a coworker, friend, or family member. Pick one shared goal (such as maybe both you want to organize your week better or create content faster). Over the next week, trade prompts and results. Compare notes. Laugh at the strange AI outputs. You will be amazed at how much faster you both learn when you are not doing it alone.

PROMPTS YOU CAN SHARE WITH OTHERS

Here are a few "shareable" prompts designed for group collaboration:

- **Team brainstorming**

    o *Generate 20 creative marketing ideas for a local coffee shop. Categorize them into: social media, in-store promotions, and community events.*

- **Improving workflows together**

    o *Suggest three automation workflows that would save time for a small team managing projects in [your tool of choice, e.g., Trello, Asana, or Notion].*

- **Cross-learning**

    o *Summarize this 5-page report in under 200 words for a nontechnical audience. Then create 2 versions: 1 for executives, 1 for staff.*

- **Creative collaboration**

    o *Draft a 5-day content plan for LinkedIn posts that 2 people can cowrite, alternating days—one with a serious tone, one playful.*

- **Feedback practice**

    o *Act as a peer reviewer. Give constructive feedback on this paragraph in 3 bullet points: clarity, tone, and engagement.*

Prompts like these are simple, but they invite everyone involved to participate, iterate, and build something together. Every time someone shares a discovery, it saves others from reinventing the wheel. Every shared success story builds confidence in those who are hesitant to try. When one person figures out how to save 10 minutes and teaches 5 others, that is nearly an hour collectively saved. Multiply that across teams, classrooms, and communities, and suddenly, AI is not making individuals more efficient; it is making everyone a little more empowered.

ACTIVITY: YOUR AI GROWTH MAP

Think of this as a personal blueprint, a visual tool that will help show you where AI already fits into your life, where you would like it to grow, and what areas you would like to keep completely human. The purpose is to give yourself a picture of your evolving relationship with AI. This exercise will help you understand this and give you a better overview of where you stand. Use it as a living document and change the information as your skills evolve, you become more comfortable using AI, and you have developed more experience.

Step 1: Divide your life into areas

On a new document, draw three circles or columns. Label them 1. Work/

school, 2. Personal life, 3. Creative/exploration. You can add more categories if you want, such as Health, Finances, and Family, but remember to keep it manageable.

Step 2: Map what you already do

In each category, jot down how you have used AI so far. Maybe in Work/School, you have asked AI to outline a report. In Personal Life, you have generated party theme ideas. In Creative/Exploration, you might have dabbled with image generation. This is your starting point, and seeing it on paper helps you realize what you have already begun.

Step 3: Add growth goals

Now, think about where you would like AI to help more. Ask yourself

- *Could AI help me save more time on routine tasks?*

- *Could it help me brainstorm or polish communication?*

- *Could AI help me organize schedules, track habits, or automate chores?*

- *Could AI help me learn a new skill, write stories, or make music?*

Add one or two goals per category.

Step 4: Mark the human zone

Just as important as deciding where to use AI is deciding where not to. Draw a small section for tasks you want to keep fully human: having deep conversations, nurturing relationships, or making final decisions on important matters. Marking this boundary helps you stay intentional.

Step 5: Create a visual version

If you enjoy visuals, you can sketch arrows, timelines, or even use Canva or Miro to make a flowchart, or you can ask a chatbot: *Help me create a visual growth map that shows how AI supports my work, personal life, and creative goals. Suggest categories, labels, and icons.* You will be surprised at how quickly AI can help you design your own road map.

As I said before, this map is not static, and you should revisit it monthly or quarterly as your needs and the technology evolve. By putting your journey on paper, you turn vague intentions into a tangible plan, and over time, you will see how far you have come. What started as small wins has become a sustainable practice that feels uniquely yours.

JOURNAL PROMPT

As you reach the final pages of this book, take a deep breath and pause. You have made it through concepts, stories, tools, prompts, and countless ideas. However, the most important part is that you have made it through your own transformation. When you first began reading, you may have approached AI with curiosity, skepticism, or even a touch of anxiety. Now, after experimenting, learning, and creating, you have built something even more valuable than knowledge: *confidence.*

The final journal is not just about reflection, but about finishing the setting of the foundation for your next chapter. You do not need to structure it perfectly. Just write what comes naturally. This space is yours; private, unedited, and honest.

How has my view of AI changed since I started this book?

When you first encountered AI, what did you feel? Were you skeptical, wondering whether it was overhyped, maybe even a little unsettling? Were you curious but unsure where to start, intimidated by unfamiliar tools or jargon? Or perhaps you were simply too busy to imagine how something so technical could fit into your life.

Now, after dozens of examples, exercises, and experiments, pause and ask yourself: *What is different?* Maybe AI no longer feels like a black box. Perhaps you have realized it does not require coding skills or expensive software, just curiosity and clear questions. Write about that shift and capture your own evolution.

What is the one habit I will commit to after finishing this book?

Maybe you will decide to ask for help with one small task each day, such as an email, a summary, or a to-do list. Maybe it will be spending 30 minutes every week exploring a new AI tool or prompt page, or sharing one AI win or useful prompt with a friend. You might keep a prompt journal where you collect the ideas and experiments that work best for you. Perhaps you will use AI to reflect on your progress each month, asking it to summarize your goals, lessons, or areas for growth.

Write it down like a promise to yourself. By pausing to reflect and committing to one simple habit, you're doing something powerful: You're making AI personal. You're saying, "This is how I will make it work for *me*."

As you write your journal entry, remember that technology will keep changing, and you will need to stay updated with its evolution. In the future, it will be your curiosity, creativity, and interaction with it that give it meaning. The goal is to stay open to keep learning, questioning, and adapting.

As you finish this reflection, take a quiet moment. Look at what you have written. It might be a paragraph, a page, or several; it does not matter. What matters is that it is yours, and it is also the beginning of your next chapter, the part of your story with AI that will be written in your own words, at your own pace. Keep going forward, refining, experimenting, and sharing your skills for years to come.

LEARNING AI IS
A LIFELONG SKILL

WHEN YOU FIRST OPENED THIS BOOK, AI MIGHT HAVE FELT like something distant: a mysterious technology reserved for tech experts, researchers, or massive corporations. But now, you have seen that it's not only accessible, but also deeply human. You have learned how to approach AI with curiosity, not fear. You have explored how it can simplify your work, spark creativity, save time, and open up new possibilities that once seemed out of reach.

However, perhaps the most important thing you have realized is that learning AI is not a one-time achievement. It is a lifelong skill that will continue to evolve alongside you, just as the technology itself evolves.

Think of AI as a companion on your journey. Some days, it is your assistant, helping you organize, summarize, or plan. Other days, it is your creative partner, your study buddy, or your business strategist. Finally, on occasion, it will be the quiet collaborator who pushes you to think differently, to ask better questions, and to refine your ideas. Like any long-term relationship, it grows richer the more thoughtfully you engage with it.

What makes a confident AI user is not technical mastery, it's mindset. A mindset that values learning over perfection, curiosity over certainty, and adaptability over fear. Each time you prompt an AI, you are practicing a new language, one of creativity and critical thinking. You are teaching yourself to think in patterns, possibilities, and questions. The goal is not to get everything right, but to explore what might be possible.

You do not need to know everything about how AI works to benefit from it. What matters is how you use it, how you question it, and how you guide it with your own wisdom and ethics. The best AI users will be those who never stop experimenting, reflecting, and sharing what they learn.

A lifelong AI skill set is not about chasing every new tool or app. It is about sustainable practice. Maybe that means dedicating a few minutes a day to try a new prompt, reflect on a result, or tweak your workflow. Maybe it's a weekly review of how AI has supported you, or a monthly goal to explore a new tool. Small, consistent actions compound over time. Wheat starts as a five-minute curiosity, and can evolve into a habit that transforms your professional and creative life. Think of it like exercise or journaling: The benefits come from showing up, not perfection.

AI is more than a collection of algorithms and interfaces; it is a reflection of us. Every prompt you write, every question you ask, and every choice you make shapes how AI will evolve not just for you, but for everyone. That is why your participation matters. By engaging thoughtfully and ethically, you help ensure AI grows as a tool that amplifies what makes us human: creativity, empathy, and insight, rather than replacing them. The future of AI is not being written by machines; it is being written by people like you who understand that technology is only as powerful as the values guiding it.

Finally, if this book has helped you see AI differently, share it. Recommend it to someone who is curious but hesitant, a friend who feels it "is too late to start," or to a colleague who could use an extra pair of digital hands. The more people who learn to use AI thoughtfully, the stronger and more creative our shared future becomes.

Keep learning. Keep questioning. Keep experimenting. You do not have to master AI; you just have to stay curious. Learning AI is not a destination; it is a lifelong skill.

Thank you for walking this journey with me. Writing this book reminded me that technology, at its best, is not about replacing what we do but about reminding us what we are capable of. Every time you open an AI tool, you are not just learning how machines think; you are rediscovering how *you* think.

So, keep exploring, keep teaching, and keep leading with curiosity. The world does not need more people who fear AI; it needs more people who use it wisely, creatively, and with heart. Here is to your next chapter, the one you will write alongside AI.

REFERENCES

AI hallucinations: ChatGPT created a fake child murderer. (2025, March 20). Noyb. https://noyb.eu/en/ai-hallucinations-chatgpt-created-fake-child-murderer

Biased health data is a mirror of society. (2025). University of Birmingham. https://www.birmingham.ac.uk/news/2025/biased-health-data-is-a-mirror-of-society

Hardesty, L. (2018, February 11). *Study finds gender and skin-type bias in commercial artificial-intelligence systems.* MIT News. https://news.mit.edu/2018/study-finds-gender-skin-type-bias-artificial-intelligence-systems-0212

Just in: email dominates business communication but kills productivity. (2021, January 8). *Mail Manager.* https://blog.mailmanager.com/blog/new-research-reveals-email-dominates-business-communication-but-poor-processes-kill-productivity-and-frustrate-employees

Probstein, S. (2019, December 17). Reality check: Still spending more time gathering instead of analyzing. *Forbes.* https://www.forbes.com/councils/forbestechcouncil/2019/12/17/reality-check-still-spending-more-time-gathering-instead-of-analyzing/

Steve Jobs' 2005 Stanford commencement address. (2023, August 15). *Tristan Ahumada.* https://www.tristanahumada.com/blog/steve-jobs-2005-stanford-commencement-address

Weiser, B., & Schweber, N. (2023, June 8). Lawyer who used ChatGPT faces penalty for made up citations. *The New York Times.* https://www.nytimes.com/2023/06/08/nyregion/lawyer-chatgpt-sanctions.html

www.ingramcontent.com/pod-product-compliance
Lightning Source LLC
Chambersburg PA
CBHW071159210326
41597CB00016B/1599